南京水利科学研究院出版基金资助

联合国教科文组织
政府间水文计划第九阶段战略规划
（2022—2029年）

——科学构筑变化环境下水安全世界

（汉英对照）

王宗志　余钟波　林祚顶◎主编

河海大学出版社
HOHAI UNIVERSITY PRESS
·南京·

图书在版编目(CIP)数据

联合国教科文组织政府间水文计划第九阶段战略规划：2022—2029年：科学构筑变化环境下水安全世界：汉英对照 / 王宗志，余钟波，林祚顶主编. -- 南京：河海大学出版社，2023. 12

ISBN 978-7-5630-8143-1

Ⅰ. ①联… Ⅱ. ①王… ②余… ③林… Ⅲ. ①联合国教科文组织－水资源管理－安全管理－研究－2022-2029－汉、英 Ⅳ. ①TV213.4

中国国家版本馆CIP数据核字（2023）第175227号

书　　名	**联合国教科文组织政府间水文计划第九阶段战略规划(2022–2029年)** **——科学构筑变化环境下水安全世界(汉英对照)** LIANHEGUO JIAOKEWEN ZUZHI ZHENGFUJIAN SHUIWEN JIHUA DI-JIU JIEDUAN ZHANLüE GUIHUA (2022-2029 NIAN) KEXUE GOUZHU BIANHUA HUANJING XIA SHUIANQUAN SHIJIE (HAN YING DUIZHAO)
书　　号	ISBN 978-7-5630-8143-1
责任编辑	彭志诚
特约编辑	张嘉彦
特约校对	薛艳萍
封面设计	槿容轩
出版发行	河海大学出版社
地　　址	南京市西康路1号（邮编：210098）
网　　址	http://www.hhup.com
电　　话	025-83737852（总编室）　025-83722833（营销部）
经　　销	江苏省新华发行集团有限公司
排　　版	南京布克文化发展有限公司
印　　刷	广东虎彩云印刷有限公司
开　　本	880mm×1230mm　1/16
印　　张	6.75
字　　数	200千字
版　　次	2023年12月第1版
印　　次	2023年12月第1次印刷
定　　价	50.00元

编 译 组

主　编：王宗志　余钟波　林祚顶

译　者：张　蓉　白　莹　池欣阳　彭　辉

王文琪　汤刘山　张玲玲　杜慧华

王　坤　熊珊珊　吴梦莹

审　校：孙　凤

审　定：余达征　Alexander Otte

序言

政府间水文计划（Intergovernmental Hydrological Program, IHP）是联合国教科文组织（United Nations Educational, Scientific and Cultural Organization, UNESCO）框架下专注水科学、水教育和水文化的政府间计划，起源于1965开始的国际水文十年（International Hydrological Decade- IHD，1965-1975），执行过程中发现10年时间难以解决水科学及其实际应用的大量问题，故UNESCO在1972年召开的第17届大会上决定IHD结束时实施IHP。成立IHP旨在促进水文国际合作，探索水领域跨学科研究，挖掘水的社会意义，解决重大水资源问题和与之相关的经济社会发展问题。IHP已经实施了八个阶段计划，目前正在实施IHP-IX，每个阶段设置不同的主题，透过“这些主题及其研究项目”之管，可窥“国际水文发展趋势”之豹。

IHP-I（1975-1980）的主题是“水科学中的水文研究方法、培训和教育（Methodologies for hydrological studies and training and education in the water sciences）”。沿袭了IHD的目标任务，从以自然界水循环、水分布研究，水文资料收集等为主要目的，开始注重人类活动影响、水资源与自然环境之间关系的研究，其间成立了水文循环影响委员会，环境水文学逐渐形成。IHP-II（1981-1983）的主题是“理性水资源管理的水文学和科学基础（Hydrology and the Scientific Bases for Rational Water Resources Management）”，在以往继续强调人类活动影响的基础上，更加重视水资源调查、评价等相关研究。IHP-III（1984-1989）的主题是“水文学及为经济、社会发展而合理管理水资源的科学基础（Hydrology and the Scuebtufuc Bases for Rational Mabagenebt of Water Resources for Economic and Social Development）”。这个阶段继续把水文科学作为重点，并把计划内容扩大到水资源合理管理，更多重视多学科共同研究水文学与水资源问题，其间提出了全球变化对水资源的影响等问题。IHP-IV（1990-1995）的主题是“变化环境中的水文和水资源可持续发展（Hydrology and Water Resources Sustainable Development in a Changing Environment）”，首次提到水资源可持续发展、生态水文学，以及环境变迁中的水文问题等新概念。IHP-V（1996-2001）的主题是“脆弱环境中水文和水资源发展（Hydrology and Water Resources Development in a Vulnerable Environment）”。在这一阶段，生态水文学得到了迅速发展，同时开始重视全球变化对环境脆弱性，以及风险中的地下水资源影响等内容。IHP-VI（2002-2007）的主题是“水的相互作用：处于风险和社会挑战中的系统（Water Interactions: Systems at Risk and Social Challenges）”。重点考虑来自地表水与地下水、大气与陆地、淡水与咸水、全球变化与流域系统、水质与水量、水体和生态系统、科学与政治、水与文化等八个方面新挑战问题，着重突出水与社会方面的研究，以及水灾害和恶化环境下的人类安全等，使人们了解更多水方面的内容。IHP-VII（2008-2013）的主题是“水的相互依赖与作用：来自各方面压力的系统和社会响应（Water Dependencies: Systems under Stress and Societal Response）”，核心内容包括全球变化、流域与浅层地下水、管理和社会经济、生态水文学与环境可持续性，水质、人类健康和食物安全。IHP-VIII（2014-2021）的主题是“水安全：应对地方、区域和全球挑战（Water Security: Responses to Local regional and Global Challenges）”，其主要目的是通过促进信息和经验转化，来满足地方和区域对全球变化适应工具的需求，并加强能力建设以满足当今全球水资源挑战所带来的挑战，从而将科学转化为行动。正在实施的IHP-IX（2022-2029），其主题是“科学构筑变化环境下水安全世界（Science for water secure world in a changing environment）”。全球水文科学家酝酿IHP-IX内容时，世界处于异常复杂的环境中。例如支撑生命的自然系统受到了人们生活与消费方式变化、人口增长、城镇化和气候变化，以及这些变化对经济社会可持续发展所需的淡水供应、极端气候事件综合效应等重大挑战影响。更为糟糕的是，全球新冠肺炎疫情爆发，人类互动交流、应对灾害的能力受到阻碍，进一步唤醒了全人类迅速开展合作、完全透明分享数据和经验教训、提出全球水安全解决方案的共识。

IHP的领导机构是政府间理事会，由UNESCO理事国组成。1979年12月我国成立IHP中国国家委员会，设在水利部，负责制订、参加IHP的我国计划，并促其实施。我国通过近50年的努力参与，特别是水文发展所取得的辉煌成就，得到了IHP各成员国的广泛认同。2021年河海大学余钟波教授当选IHP理事会主席（这是我国专家首次当选这一重要职务），2023年又顺利当选理事会特别督导/副主席，这标志着IHP中国国家委员会的作用和地位不断巩固提高。考虑到以往（IHP I-VIII）战略规划没有正式出版中文版，可能会在一定程度上影响我国政府、学者、公众参与联合国涉水事务的深度和广度。为此，IHP中国国家委员会秘书处办公室（设在南京水利科学研究院），在水利部国际合作与科技司、水文司的具体指导下，2022年初组织翻译IHP-IX战略规划。经过近2年的翻译、校核、审定，由UNESCO与河海大学出版社联合出版、UNESCO网站发布的IHP-IX战略规划即将面世。希望此举，能为提升我国在涉水领域的国际话语权和影响力、助力我国水文事业走向世界、分享中国治水经验、拓宽水领域国际合作略尽绵薄之力。

在IHP-IX战略规划中文版翻译、审校与审定过程中，得到了水利部国际合作与科技司、水文司、南京水利科学研究院等单位领导的大力支持，UNESCO水科学局局长、IHP秘书长Abou Amani与Alexander Otte先生为与UNESCO联合出版做出了大量的协调与校验工作。在此，编译组谨向他们表示衷心感谢。特别感谢国家自然科学基金项目（编号U2240223）与南京水利科学研究院出版基金的资助。

需指出的是，囿于成稿仓促，编译组水平有限，与“信、达、雅”的要求相距甚远，殷切希望同行专家和读者朋友给予批评指正。

王宗志

2023年12月于南京清凉山麓

目录

CONTENTS

说明…… 001
前言…… 005
第一章　全球水格局：挑战和机遇…… 009
第一节　实现可持续发展目标，落实其他涉水国际议程 …… 010
第二节　推动水科学的跨学科发展 …… 012
第三节　加强水教育，支撑当今和未来可持续发展 …… 014
第四节　弥合数据与知识之间的鸿沟 …… 016
第五节　努力实现可持续和包容的水资源综合管理 …… 016
第六节　科学支撑水治理 …… 018
第七节　实现变化世界中的水安全 …… 018

第二章　政府间水文计划第九阶段战略规划的制定背景和战略目标…… 021
第一节　联合国教科文组织和政府间水文计划的优势 …… 022
第二节　确保变化环境下政府间水文计划战略的连续性 …… 022
第三节　政府间水文计划的战略目标 …… 024
一、政府间水文计划的愿景 …… 024
二、政府间水文计划第九阶段战略规划的使命 …… 024

第三章　政府间水文计划第九阶段战略规划预期成果和优先领域…… 027
第一节　成果链与变革理论 …… 030
第二节　优先领域 …… 032
一、优先领域 1：科研与创新 …… 034
二、优先领域 2：第四次工业革命中可持续发展水教育 …… 044

三、优先领域 3：弥合数据与知识之间的鸿沟 …… 052
四、优先领域 4：全球变化条件下的水资源综合管理 …… 058
五、优先领域 5：基于科学的缓解性、适应性和韧性水治理 …… 068

第四章　政府间水文计划第九阶段战略规划实施手段和实施路径 …… 077
第一节　沟通和外联 …… 078
第二节　实施路径 …… 078

附件 …… 083
1　术语表 …… 084
2　伙伴清单（暂定） …… 088
3　现代水文学的主要问题 …… 090
4　政府间水文计划旗舰倡议清单 …… 094
5　变化理论图 …… 098

说明

DESCRIPTION

1 政府间水文计划第九阶段战略规划时间跨度为2022—2029年，确定了关键的水优先领域，以支持成员国落实《2030年可持续发展议程》、实现可持续发展目标，特别是涉水可持续发展目标以及其他涉水全球议程，例如《巴黎气候变化协定》、《仙台减少灾害风险框架》和《新城市议程》。

2 本次战略规划编制过程体现了高度的参与性。在连续磋商阶段，收集了区域专家、政府间水文计划执行局和理事国、联合国教科文组织涉水机构、伙伴组织和联合国机构的实质性的、有用的意见。

3 以下三个相互关联的文件将指导政府间水文计划第九阶段战略规划的实施：1）本战略规划文本（用于确定成员国涉水优先事项）；2）业务实施计划；3）财务战略计划。后两个文件将在后面阶段详细说明，将用于通过拟定行动方案和相关指标跟踪战略规划实施的进展情况。

1 The Strategic Plan for the ninth phase of the Intergovernmental Hydrological Programme (IHP-IX) covering 2022—2029 identifies key water priority areas to support Members States to achieve the 2030 Agenda and the Sustainable Development Goals (SDGs), especially water related SDGs and other water-related global agendas, such as the Paris Agreement on Climate Change, Sendai Framework on Disaster Risk Reduction (DRR) and the New Urban Agenda (NUA).

2 The process of preparing the Strategic Plan has been highly participatory, collecting in successive consultation stages the inputs of regional experts, the IHP Bureau and Council members, the UNESCO Water Family, partner organizations and UN agencies, whose observations were substantial and useful.

3 The implementation of the ninth phase of the IHP (henceforth IHP-IX) will be guided by three interrelated documents: i) a Strategic Plan, presented herein, identifying water-related priorities for Member States, ii) an Operational Implementation Plan, and iii) a Financing Strategy, the last two documents to be elaborated at a later stage, which will be used to track the progress in implementing the Strategic Plan through proposed actions and related indicators.

前言

PREAMBLE

4 联合国教科文组织政府间水文计划成立于1975年，是一项长期计划，每8年为一个执行阶段。其重点经历了从单一学科模式到多学科并举的深刻转变，旨在通过支持科学研究和教育计划，推动水文学知识的发展。自二十一世纪初以来，随着社会科学元素的增加，包括公民科学投入的质量和数量的提升，政府间水文计划已经演变成一项真正的跨学科事业。此项进步得益于以下认识：世界水问题的解决方案，不仅涉及技术、工程或自然科学，而且深受人文和社会文化因素的影响，而社会科学在其中发挥着日益重要的作用。

5 政府间水文计划是一个政府间合作计划，旨在通过提升对水的科学认识、提高技术能力和加强教育，应对国家、区域和全球水挑战，建设一个可持续的韧性社会。为了实现这些目标，创新发展科学技术，并根据迄今获得的各种数据、知识、经验和客观事实来生成综合科学知识，广泛分享、使用这些知识，这些都很重要。此外，政府间水文计划旨在开发负责这些工作的人力资源。因此，只有建立一些使所有利益相关方都能参与的机制，才可能支持基于科学技术作出决策的治理。

6 我们生活在一个空前的风险与未来的巨大机遇并存的时代。许多科学家认为，支撑生命的自然系统受到了诸如生活方式和消费模式变化、人口增长、城镇化和气候变化，以及这些变化对水文行为、人类和可持续发展所需的淡水供应、极端气候事件综合效应等当代最大挑战的影响。像全球新冠疫情这样的灾害造成了额外的边界条件限制，制约了人类互动以及应对灾害导致的同步和/或连锁影响的能力。此例证明了科学、研究和技术的极端重要性；为了全球社会的集体利益，各国需要迅速开展合作，并完全透明地分享数据和经验教训。

7 环境变化与人类活动相互交织，其变化速度日益加快，呼吁我们更好地理解水文学。在制定水资源综合管理方案时，需要考虑人类活动与水系统的相互作用。政府间水文计划第九阶段战略规划继续为扩大国际科学界合作提供平台和场所，从而有助于解决水文学中许多未解决的问题[1]。

8 本战略规划的目的是为2022—2029年期间的政府间水文计划提炼出世界公认的战略焦点。第九阶段战略规划提出了解决复杂背景下水安全问题的跨学科方法体系。其中所提出的方法和优先事项符合联合国教科文组织在科学和教育领域的主要任务，旨在当成员国面临全球涉水挑战时，响应他们的需求，并支持他们利用科技进步来应对挑战。

1 Blöschl，G. 等人（2019）；水文学的二十三个未解决问题——水文界观点，《水文科学杂志》，第64卷第10期，第1141-1158页。

4 The UNESCO Intergovernmental Hydrological Programme (IHP), founded in 1975, is a long-term programme executed in successive eight-year phases. Its programmatic focus has gone through a profound transformation from a single discipline mode, to a multi-disciplinary undertaking, aimed at advancing hydrological knowledge through supporting scientific research and educational programmes. Ever since the early 2000s, with the increased presence of social science components, including growth in the quality and quantity of citizen science inputs, IHP has been evolving into a truly transdisciplinary undertaking. This progress has capitalized on the recognition that solutions to the world's water-related problems are not just technical, engineering or natural science issues, but have strong human and socio- cultural dimensions, where social sciences play an increasingly important role.

5 The IHP is an intergovernmental cooperation programme aimed at addressing national, regional and global water challenges and building a sustainable and resilient society by expanding the scientific understanding of water, improving technical capabilities, and enhancing education. To attain these aims, it is important to innovatively develop science and technology, generate integrated scientific knowledge based on various data, knowledge, experience and objective facts obtained so far and share this knowledge widely, and implement it. Furthermore, IHP aims at developing human resources who will be responsible for these endeavours. As a result, it will become possible to support governance whereby decisions are made based on science and technology by building mechanisms enabling all stakeholders to participate.

6 We live in a time of unprecedented risks, but also of great opportunities for the future of our planet. Natural systems that support life are affected by what many scientists consider the supreme challenges of our time such as changing lifestyles and consumption patterns, population increase, urbanization, and climate change, and their impact on hydrological behaviours, the availability of freshwater for human consumption and for sustainable development and the combined effects of extreme climatic events. Disasters such as the COVID-19 global pandemic caused additional limitations to boundary conditions, restricting the capability of humans to interact and deal with synchronous and/or cascading impacts of disasters. This example has demonstrated the utmost importance of science, research and technology as well as the need for rapid cooperation and full transparency regarding the sharing of data and lessons learned for the collective benefit of the global community.

7 The ever-increasing pace of environmental changes intertwined with human behaviours calls for better understanding of hydrology. The interaction between human activities and water systems needs to be considered to develop scenarios for Integrated Water Resources Management (IWRM). IHP-IX continues to offer a platform and venue to extend cooperation within the international scientific community, and thus contribute to addressing many unsolved problems in hydrology[1].

8 The purpose of this strategic plan is to outline a compelling and strategic focus for the Intergovernmental Hydrological Programme for the 2022—2029 period. The ninth phase represents a methodological response towards transdisciplinarity aimed to generate solutions for a water secure world in a complex context. The approach and prioritization presented aligns with UNESCO's principal mandates in the Sciences and Education, and aims to be responsive to the needs of Member States and support them to capitalize on scientific and technological advances as they face water-related global challenges.

1 Blöschl, G. et al. (2019) Twenty-three unsolved problems in hydrology (UPH) – a community perspective, Hydrological Sciences Journal, 64:10, 1141-1158.

第一章
全球水格局：挑战和机遇

GLOBAL WATER LANDSCAPE:
CHALLENGES AND OPPORTUNITIES

9 虽然在大多数情况下水是可再生资源，但是人类活动导致水资源日益短缺，大都市、生产中心、农业纬度地区、干旱和半干旱地区的水资源尤为短缺。经济发展和人口增长对水平衡和淡水质量产生影响，意味着专家和所有用水户需要深化和扩展水文学和水资源管理的综合知识。此外，气候变化加剧了涉水挑战的严重性，增强了将水议程推向全球社会前沿的紧迫性。

10 我们当前面临的挑战是相互关联的，无法像往常一样采用单一部门的应对方式。因此，通过跨学科的科学研究，开展涉水可持续管理教育和培训以优化用水，是构筑可持续发展体制和水治理的必要基石，以便应对全球变化、实现水安全。

11 淡水是人类生命、健康、食物安全和生物多样性的基础。全世界数十余亿人口面临着缺水、水质欠佳、卫生条件差以及涉水灾害的挑战。到 2030 年，世界近一半人口将生活在严重缺水地区（《联合国世界水发展报告》，2019 年）。由于气候变化和人口增长，涉水风险将进一步加剧。此外，移民正在加剧水资源的压力。事实上，截至 2016 年底全球有超过 6 500 万人被迫流离失所（联合国难民署，2017 年），遗憾的是，这一趋势仍在加剧。另外，全球各国从农村到城市的移民人口正在普遍增涨，给城市服务带来了更多压力。目前，超过 50% 的世界人口居住在城市。

12 然而，仅仅认识和理解水管理者和所有利益相关方所面临的问题是不够的，重要的是为第九阶段及以后阶段的政府间水文计划明确当前和未来的潜在机遇。必须识别有效的解决方案，并将其作为与科学、教育和技术相关的战略和对策的一部分，纳入政府间水文计划第九阶段战略规划。

13 同样重要的是，要区分在 2022—2029 年期间实施跨学科计划所需各类角色及其作用。2019 年，年龄在 15—24 岁之间的人口大约 12 亿，约占世界人口的六分之一；到 2030 年，预计这一数字将增长 7%[2]。因此，鼓励青年参与政府间水文计划对于培养致力于发展水文化、实现水安全和可持续发展目标（水管理）的未来一代领导人至关重要。联合国教科文组织的青年业务战略将为年轻专家参与政府间水文计划奠定基础。同样，作为提升水科学、水文化以及改善水管理、水治理的变革推动者，妇女和女孩发挥着关键作用。考虑到土著群体了解关于水问题的祖传知识，他们的作用也将至关重要。

第一节　实现可持续发展目标，落实其他涉水国际议程

14 成员国面临的一项主要挑战是实现《2030 年可持续发展议程》的可持续发展目标。尽管各成员国已经为此付出了努力并投入了资源，但严重偏离了 2023 年实现可持续发展目标 6（人人享有水和卫生设施）的轨道。《2020 年联合国世界水发展报告》强调，水是“气候连接器”，它允许在涉及可持续发展、气候变化以及减少灾害风险等领域的大多数目标上开展更大的合作和协调。

15 可以肯定的是，大多数全球议程与水直接相关，而其他议程与水间接相关，且在实现可持续发展目标 6 方面的任何改进都会对其他议程产生附带影响。正在进行的联合国改革进程以支持成员国实现可持续发展目标为导向，为政府间水文计划通过其国家委员会、教席和中心更多地从国家和区域层面上参与提供了更大的机会。

2 联合国经济和社会事务部（2019 年）。

9 Although water is mostly a renewable resource, it is increasingly scarce due to human activity, particularly in large metropoles, production centres, agricultural latitudes, arid and semi-arid regions. The impact of economic and demographic growth on the water balance and the quality of freshwater means it is necessary to deepen and expand our comprehensive knowledge of hydrology and management of water, not only among experts, but also among all users. Furthermore, climate change exacerbates the gravity of water-related challenges and increases the urgency of bringing the water agenda to the forefront of the global community.

10 The challenges we are currently facing are interconnected and cannot be met if we continue a business as usual, sectoral-silo approach. That is why optimizing the use of water through transdisciplinary scientific research, together with education and training for its sustainable management, constitute necessary keystones for sustainable institutional development and water governance for global change and water security.

11 Freshwater is essential to human life, health, food security and biodiversity, challenges of water scarcity and quality, poor sanitation and water-related disasters confront billions worldwide. Almost half of the world's population will be living in high water stress areas by 2030 (WWDR, 2019). Water-related risks will further increase because of climate change and demographic growth. Furthermore, human migrations are putting pressure on water resources. Indeed, more than 65 million people were involuntarily displaced as of the end of 2016 (UNHCR, 2017) and unfortunately the trend has continued to increase. Also, migration within countries due to relocation of people from rural to urban areas is increasing globally, putting more pressure on urban services. Currently more than 50% of the world's population lives in cities.

12 However, it is not sufficient to simply recognize and understand the problems water managers and overall stakeholders are facing; it is important to ascertain the present and potential future opportunities that are available to project IHP into its ninth phase and beyond. Effective solutions must be identified and included in the programmes proposed for IHP-IX, as part of scientific, education and technology related strategies and responses.

13 It is equally important to distinguish the various players and their role in the implementation of the transdisciplinary programme envisioned for the 2022—2029 period. In 2019, there were about 1.2 billion persons between the ages of 15 and 24, about one in every six persons in the world. This number, which is projected to grow by 7 percent by 2030[2], makes youth engagement essential for building a generation of future leaders committed to an evolved water culture, water security and achieving the SDGs (water stewardship). UNESCO's Operational Strategy on Youth will provide the basis upon which the involvement of young experts will be founded. Similarly, women and girls play key roles as agents of change for improved water science, culture and better water management and governance. The role of indigenous groups, considering their ancestral knowledge on water issues will also be essential.

Meeting the SDGs and other water-related International Agendas

14 A major challenge confronting Members States is meeting the Sustainable Development Goals (SDGs) that comprise the UN 2030 Agenda for Sustainable Development. Despite the efforts and resources dedicated to this task, the SDGs are off track with an alarming trend for SDG 6 (Water and Sanitation for all). The United Nations World Water Development Report (WWDR) 2020 emphasizes that water is the "climate connector" that allows for greater collaboration and coordination across most targets for sustainable development, climate change, and disaster risk reduction.

15 Certainly, the majority of the global agenda are directly related to water while others are connected indirectly and any improvement in the achievement of SDG 6 results in having secondary effects on them. The on-going process of UN reform with SDGs country-oriented support provides a greater opportunity for IHP through its national committees, Chairs and Centres for more engagement at country and regional levels.

2 United Nations Department of Economic and Social Affairs (UNDESA 2019)

16 政府间水文计划第九阶段战略规划内容旨在通过加强科学知识和数据可用性以及促进明智决策，以便最大程度地支持成员国实现可持续发展目标 6 及其相关的“全球加速框架”、联合国《十年水行动宣言》（2018—2028 年）以及其他涉水目标和指标。此外，与联合国欧洲经委会一起作为可持续发展目标 6.5.2（水资源合作业务安排的跨界流域面积比例）的共同监管方，政府间水文计划获得了一个独特的机会，在可持续发展目标 6 和国际公约框架内，在帮助各国明确和实施跨界含水层管理和保护方面将发挥重要作用，以便达到与改善水资源合作业务安排有关的必要基准。

17 迄今为止，水管理和水工程的重点是提供农业、工业、航运和生活用水。然而，水资源大量开发、流域改造和气候变化放大了水文过程的随机性，并且增大了洪水和干旱的强度和频率。这将反过来进一步降低人均水资源占有率。

18 从广义上讲，实现可持续发展目标意味着承认人类与生物圈之间的复杂关系，特别是承认水作为生物生产力、生物多样性以及营养循环、所有基本生命支持过程关键驱动力的作用，因此，迫切需要协调水资源需求与加强水管理之间的关系。因此，水管理面临的最重要挑战是如何增加水量、提高水质，同时增强生物多样性、生态系统的社会服务功能以及应对影响的复原力。应对这些挑战需要一套整体分析方法，将创新的、基于自然的生态水文学解决方案与流域层面的系统解决方案相结合。一方面，理解作为水的接受者和生产者的水生态系统之间的相互作用；另一方面，通过有关水和可持续发展方面的文化和教育活动，促进社会参与水管理。水文化深刻体现在社会行为模式中，它描述了人类的感知、行动、效率和效力。

政府间水文计划第九阶段战略规划定位是：在全球涉水政策背景下，与其他倡议保持一致，并帮助实现这些倡议目标。其中，适用于署名成员国的重要框架为：可持续发展目标框架及《2030 年可持续发展议程》，特别包括有关确保所有人享有水和卫生设施及其可持续管理的可持续发展目标 6，以及它与其他可持续发展目标的联结作用；相关的“可持续发展高级别政治论坛”以及对可持续发展目标进展的最新监控；“可持续发展目标 6：全球加速框架”；《联合国气候变化框架公约》下的《巴黎协定》；《仙台减少灾害风险框架》；《亚的斯亚贝巴发展筹资行动议程》；《新城市议程》；人类享有安全饮用水和卫生设施的《人权框架》（联合国大会第 A/RES/64/292 号和第 A/RES/70/169 号决议）；《全球粮食安全和营养战略框架》；《跨界水道和国际湖泊保护和利用公约》（赫尔辛基，1992 年）；《国际水道非航行使用法公约》（纽约，1997 年）；以及《跨界含水层法》（联合国大会第 A/RES/68/118 号决议）。其他重要框架包括联合国大会《水行动十年宣言》（2018—2028 年）和《实现 2030 年可持续发展目标十年行动计划》、“联合国生态系统恢复十年（2021—2030 年）”行动、“海洋科学促进可持续发展十年（2021—2030 年）”计划、全球适应委员会的“行动年”，以及《小岛屿发展中国家快速行动方式（萨摩亚途径）》的成果文件。

第二节　推动水科学的跨学科发展

19 环境变化要求更好地理解水文，而水文则是这些变化的关键因素。水文科学界最大挑战之一是在不断变化的环境中确定恰当的、及时的适应和缓解措施。有必要研究非平稳气候变化条件下水文过程的物理机制，然后研究这些条件下的适应机制。此外，需要从新的角度分析人类和水系统之间的相互作用，以便全面了解内在反馈和协同演化的过程和情景。为了应对这些挑战，政府间水文计划需要制定一个

16 The IHP-IX programmatic content is designed with the objective to maximize the support to Member States in attaining SDG 6 and its related UN SDG 6 Global Accelerator Framework, UN Water Decade for Action (2018—2028) and other water-related goals and targets by strengthening scientific knowledge, data availability and enabling informed decision-making. In addition, being the co-custodian of SDG 6.5.2 (Proportion of transboundary basin area with an operational arrangement for water cooperation) with UNECE, provides a unique opportunity to IHP to play a major role in identifying and implementing actions to help countries, within the framework of SDG 6 and international conventions, achieve the required benchmarks pertaining the improvement of operational arrangements for water cooperation related to the management and conservation of transboundary aquifers.

17 To date, water management and water engineering have been focused on water supply to agriculture, industry, navigation, and domestic use. However, intensive water exploitation, catchment modification and climate change have amplified the stochastic character of the hydrological process and both the intensity and frequency of floods and droughts. This will in turn further negatively impact the water resources per capita ratio.

18 In a broader context, attaining the SDG implies recognizing the complex relation between humans and the biosphere, and in particular the role of water as a key driver of bio-productivity, biodiversity, and nutrient cycles, all fundamental life supporting processes, thus the urgent need to harmonize the demand with enhanced water resources. Consequently, the most important challenge for water management is how to increase water resources quantity and quality, and in parallel to increase biodiversity, ecosystem services for society, and resilience to impacts (WBSR). The answer is through a holistic approach which incorporates the innovative nature-based and ecohydrological solutions methods (NBS) and catchment scale systemic solutions, based on understanding the water ecosystems interplay, as both recipients and producers of water, as well as promoting society involvement through culture and education of water and sustainability (CE). Culture is deeply embodied in societal behaviour patterns, defining perception and actions, efficiency and effectiveness.

The IHP-IX Strategic Plan is positioned within the context of the global water-related policy landscape to provide opportunities of alignment with other initiatives and in contributing to their achievement. This landscape consists of, among others, the following key frameworks (applicable for the signatory Member States): the Sustainable Development Goals (SDG) framework and its 2030 Agenda including, specifically, SDG 6 on ensuring availability and sustainable management of water and sanitation for all and its connecting role to all the other SDGs, the associated High-Level Political Forum on Sustainable Development and updated monitoring of progress towards SDG targets, the SDG6 Global Accelerator Framework, the Paris Agreement within the UN Framework Convention on Climate Change, the Sendai Framework for Disaster Risk Reduction, the Addis Ababa Action Agenda for Financing Development, the New Urban Agenda, the Human Rights Framework with reference to the human rights to safe drinking water and sanitation (UNGA Resolution A/RES/64/292 and A/RES/70/169) and the Global Strategic Framework for Food Security and Nutrition, the Convention on the Protection and Use of Transboundary Watercourses and International Lakes (Helsinki 1992), the Convention on the Law of the Non-navigational Uses of International Watercourses (New York, 1997), and the Resolution A/RES/68/118 on the Law of Transboundary Aquifers. Other important frameworks include the UNGA declaration on the Water Action Decade 2018—2028 and the Decade of Action to Deliver SDGs by 2030, the UN Decade on Ecosystem Restoration (2021—2030), the Decade of Ocean Science for Sustainable Development (2021—2030), the Global Commission on Adaptation's Year of Action, and the outcome document of the Small Island Developing States' Accelerated Modalities of Action (SAMOA) Pathway.

Advancing water science while expanding towards transdisciplinary

19 Environmental changes are and demand a better understanding of hydrology as a key factor of these changes. One of the greatest challenges for the hydrologic science community is to identify appropriate and timely adaptation and mitigation measures in a continuously changing environment. There is a need to study the physics of hydrological processes under conditions of non-stationary climate change, and then adaptation mechanisms under these conditions. Furthermore, the interaction between human and water systems needs

新愿景。政府间水文计划第九阶段战略规划将利用不断增强的计算能力、新的监测技术、强大的建模能力、信息共享新机遇、非传统数据源以及广泛的国际和跨学科合作来实现这一愿景。

20 政府间水文计划第八阶段战略规划的基本前提是对高质量科学的承诺，以便作出明智的水决策，并提高人们生活质量。政府间水文计划第九阶段战略规划将继续秉持并加强这一承诺。题为“未来即现在：科学促进可持续发展”的《2019 年全球可持续发展报告》明确强调，科学证据是设计和实施可持续发展转型的先决条件，要求成员国与科学界（如：研究协会、大学、中心）开展合作。同样，水治理高级别小组报告《每一滴水都很重要》也明确强调了循证决策在应对复杂水挑战方面的迫切性。

21 政府间水文计划第九阶段战略规划实施期间将支持科学知识的拓展和分享，其中有些与公民科学相关。这一新兴领域结合了科学家和公众的力量，以便更好地了解水循环，包括人类行为的影响。同样，鼓励开放型科学和开放型数据，使科学信息、数据和产出更具包容性、更易获取，并且在所有利益相关方（科学家、决策者和公民）的积极参与下，被更可靠地加以利用。另外，社会水文学等学科是研究人-水动态交互的跨学科领域，并展示了联合国教科文组织在开展水科学研究方面向跨学科研究的转变。此外，通过发展与用水户、私营企业家和非政府组织的伙伴关系，建立知识库和社区信任，可以扩大应用创新科学成果和采用新技术所产生的积极影响，特别是对农村地区和传统社会的积极影响。加强科学研究与合作有助于弥合数据与知识之间的鸿沟。

第三节　加强水教育，支撑当今和未来可持续发展

22 教育可从根本上改变人的行为，并为可持续水资源决策建立共识。尽管国际社会越来越认识到可持续发展教育的重要性，在第四次工业革命期间将未来可持续发展水教育作为主要内容纳入正规和非正规教育课程仍然是一个挑战。不同地区开发的各种网络、倡议、公共设施和工具尚未成功地对教育政策和实践产生重大影响。为了通过国家层面的决策和实施从体制上加强可持续发展教育，需要投入更多力量，以便验证和推广一种基于经验的教育模式。同样，需要加强可持续水文化和水管理方面的能力和公共意识。为了应对水资源短缺问题，需要对所有形式的生产和消费（从个人使用到制造和供应链）进行重大改革，因此，需要创新教育方案。

23 大力推动对水管理人员的新技术教育，已经极大地缩小了制约他们进行合格的水治理的技能差距。然而，旨在加强法律、政策和体制框架以支撑水治理的教育事业已经滞后，构成了挑战。这一现实为政府间水文计划提供了确定并开展基于科学的能力建设活动的机会，使成员国能够加强从地方到流域层级的水治理。

24 水、能源、粮食和生态系统之间的关系是实现可持续发展的关键要素。因此，应将对这一关系及其综合方法的理解纳入各级正式和非正式的教育方案。政府间水文计划第九阶段战略规划将利用“人与生物圈计划”、“国际地球科学与地质公园计划”、“地方和土著知识系统”计划以及联合国教科文组织教育部门和联合国环境规划署在“可持续消费教育”方面的努力，为未来可持续发展采取行动，推进水教育和能力拓展活动。在第九阶段，政府间水文计划将与联合国教科文组织水事机构和其他分支机构以及联合国其他机构、组织密切合作，加强参与性和跨学科的知识建设和传播。同样，政府间水文计划第九阶段战略规划将推进旨在加强和支持成员国制定和实施国际框架、机构跨部门合作的教育计划。

to be analysed from new perspectives to develop a comprehensive picture of the inherent feedbacks and coevolving processes and scenarios. To address these challenges, IHP needs to develop a new vision. IHP-IX will take advantage of the increasing computational power, and new monitoring techniques, enhanced modelling capabilities, new opportunities for sharing information, non- traditional data sources and greater international and transdisciplinary cooperation to achieve its vision.

20 The basic premise of IHP-VIII was its commitment to quality science for informed water decisions and its impact to improving quality of life. This commitment will continue and be built on during IHP-IX. The 2019 Global Sustainable Development Report: "The Future Is Now: Science for Achieving Sustainable Development" has clearly highlighted that scientific evidence is a prerequisite for designing and implementing transformations to sustainable development requiring Member States to work with the scientific community (e.g. research consortiums, universities, centres). Similarly, the High-Level Panel on Water report (Every drop counts) has also clearly stressed the crucial need for evidence-based decisions in addressing complex water challenges.

21 Among the several opportunities for expanding and sharing scientific knowledge that will be supported during IHP-IX are those that relate to citizen science. This emerging field combines the efforts of scientists and the public to better understand the water cycle, including the effects of human behaviour. Similarly, encouraging Open Science and Open Data provides an opportunity for scientific information, data and outputs to be more inclusive, more widely accessible and more reliably harnessed with the active engagement of all the stakeholders (scientists, policy-makers and citizens). Furthermore, disciplines, such as socio-hydrology, provide an interdisciplinary field studying the dynamic interactions between water and people and an opportunity to showcase UNESCO's transition towards transdisciplinarity when conducting scientific research on water. Additionally, partnerships with water users, private entrepreneurs and NGOs to build a knowledge base and the trust of communities could multiply the positive impacts of applying innovative scientific findings and employing new technologies, particularly in rural and traditional societies. Enhancing scientific research and cooperation can help bridge data and knowledge gaps.

Strengthening water education for a sustainable present and future

22 Education is still the foundation upon which behaviours can change and consensus can be built for sustainable decisions on water resources. Despite increasing international recognition of Education for Sustainable Development (ESD), mainstreaming of water education for a sustainable future during the fourth industrial revolution in formal education curricula as well as in informal education remains a challenge. Various networks, initiatives, utilities and tools, which have been developed in different regions, have not yet succeeded in having a significant impact on educational policies and practices. More efforts are needed to validate and disseminate an experience-based model for institutional strengthening of ESD through policymaking and implementation at a national level. Similarly, enhanced capacity and public awareness towards a sustainable water culture and water management are required. Coping with water scarcity will entail a major overhaul of all forms of production and consumption, from individual use to manufacturing and supply chains and consequently will require innovation in educational programmes.

23 A major push in educating water managers on new technologies has substantially reduced the skill-gap limiting adequate water governance. However, educational undertakings aimed to enhance legal, policy and institutional frameworks to support water governance have lagged, which constitutes a challenge. This reality provides IHP with an opportunity to identify and pursue science-based capacity building activities to enable Member States towards enhanced water governance from local to basin levels.

24 The water, energy, food, and ecosystem (WEFE) nexus is a key element to consider to achieve sustainable development. Therefore, the understanding of the nexus and its integrative approaches should be incorporated in educational programmes at all levels, formal and informal. IHP-IX will use the opportunity of linking with the Man and Biosphere Programme (MAB), the International Geoscience and Geopark Programme, the Local and Indigenous Knowledge Systems (LINKS) programme and also with UNESCO's Education Sector and UNEP's efforts on "Education for Sustainable Consumption" to undertake actions to advance water education

25 联合国教科文组织致力于开放教育资源工作，为提高学习和知识共享的质量以及促进全球政策对话、知识共享和能力建设提供了战略机会。

第四节　弥合数据与知识之间的鸿沟

26 信息和知识增长需要新理念和新模式，使我们能够从中获益。各成员国监测水文过程，管理、储存和分析数据与信息，以及最终开发和应用模型的能力千差万别。确保各国和各地区之间进行可靠的决策数据、信息和知识交换，仍然是一项重大挑战。此外，非常有必要完善科学组织与联合国机构、成员国以及利益相关方之间管理水数据的合作战略。联合国教科文组织完全有能力为世界气象组织、粮农组织、环境规划署全球环境监测系统、可持续发展目标综合管理信息系统等联合国涉水部门的其他数据相关倡议作出贡献。

27 尽管如此，新监测技术的开发和应用，特别是遥感数据的发展与应用，为在大范围时空尺度上观测水文过程提供了机会。同样，全球尺度水文模型为更全面地了解和绘制可用水资源分布并识别水灾害提供了振奋人心的机会。全球尺度水文模型的不确定性仍然是一个挑战，但随着全球数据集的增长，这种不确定性正在逐步降低。由此，在更大空间尺度上开展工作为应对全球水挑战提供了新途径。诸如世界气象组织全球水文测量支持设施等全球化的新倡议旨在支持终端用户获取水文气象数据以及来自各经济部门的定制服务信息。政府间水文计划第九阶段战略规划将利用这些机会，全力帮助成员国开发和获取数据和信息，为其提供一个促进数据共享和知识传递的平台。

第五节　努力实现可持续和包容的水资源综合管理

28 淡水是把社会各方面联系在一起的“那根线”。它既维持地球上的生命，也扶持可持续经济发展。健康的河流、湖泊、湿地、含水层和冰川不仅提供饮用水和维持宝贵的生态系统，而且支持世界各地的农业、水力发电和防汛抗旱。因此，流域层级可持续水资源管理必须是一项综合性的事业——它的成功需要所有用水户（民间团体、机构、公共当局以及私营部门）的参与。

29 政府间水文计划第九阶段战略规划将促进跨学科方法的运用，把水科学与经济社会各方面纳入一个模型，旨在使生态系统商品和服务产出最大化，提高各生产部门的传统经济产出。该方法支持自然科学和行为科学的融合，促进了项目的“协同创新”和“协同设计”。

30 这种模型面临诸多困难，包括建模的时空尺度和复杂性，数据的采集、质量、处理和率定，结果的解释和传播，以及公民科学投入的整合。

31 水管理是一个长期过程，需要有几十年以上的愿景。这样的愿景可通过制定国家级和次国家级水管理总体规划来实现，这些规划将考虑自然条件、技术解决方案、基于自然的解决方案和社会现实。

and capacity development activities for a sustainable future. During its ninth phase, IHP will work closely with the UNESCO Water Family, other branches of UNESCO and UN Agencies and Organizations to enhance participatory and cross-disciplinary knowledge building and dissemination. Likewise, IHP-IX will advance educational programmes aimed at strengthening and supporting Member States in the development and implementation of international frameworks and institutional cross-sectoral cooperation.

25 UNESCO is dedicated to Open Education Resources (OER). OER provide a strategic opportunity to improve the quality of learning and knowledge-sharing as well as improve policy dialogue, knowledge-sharing and capacity-building globally.

Closing the data–knowledge gap

26 The increase of information and knowledge requires new ideas and models that will allow us to profit from these. The capacity of individual Member States to monitor hydrological processes, manage, store and analyse data and information, and ultimately develop and apply models is very heterogeneous. Ensuring that data, information and knowledge exchange among states and regions is reliable so that it can be factored into decision-making is still a major challenge. Furthermore, elaboration of collaborative strategies for managing water data between scientific organizations and UN agencies, Member States and stakeholders is very much needed. UNESCO is well positioned to contribute to other data-related initiatives across the UN-Water constituency such as those of WMO, FAO, UNEP-GEMS, SDG-IMI etc.

27 Notwithstanding, the development and application of new monitoring techniques, and in particular remotely sensed data, are providing opportunities for observing hydrological processes across a wide range of scales, both temporal and spatial. Similarly, global-scale modelling is presenting stimulating opportunities for acquiring a more comprehensive understanding and mapping of water resources availability and identification of water threats. Uncertainties in global-scale modelling are still a challenge, but they are being reduced by growing global data sets. Consequently, working on larger spatial scales puts forward new ways to address global water challenges. New initiatives on global level like the WMO Global Hydrometry Support Facility (WMO HydroHub) aim at supporting access to end-users of hydrometeorological data and services from various economic sectors as tailored services. IHP-IX will build on these opportunities to expand the efforts of the Programme aiming to help Member States develop and gain access to data and information and to provide a platform to further increase data sharing and knowledge transfer.

Working towards sustainable and inclusive integrated water resources management

28 Freshwater is “the thread” that ties all aspects of society together. It supports life on Earth and it also supports sustainable economic development. Healthy rivers, lakes, wetlands, aquifers, and glaciers not only provide drinking water and maintain valuable ecosystems; they also support agriculture, hydroelectric power, flood mitigation and drought prevention around the world. Therefore, sustainable water management at the basin level must be an integrated undertaking —— to be successful it needs to involve all water users (civil society, institutions, public authorities and private sector).

29 A transdisciplinary approach, which IHP-IX will help to promote, integrates water sciences and economic and social aspects into a model, with the objective of maximizing ecosystem goods and services outputs as well as more traditional economic outputs in various productive sectors. This approach supports convergence between natural and behavioural sciences and leads to “co-innovation” and “co-design” of projects.

30 Such a model faces many difficulties in terms of temporal and spatial scales and the complexity of model formulation, data collection, quality, handling and calibration, results interpretation and dissemination, the incorporation of citizen science inputs.

31 Water management is a long-term process that requires a vision that goes beyond decades. Such a vision can be reached by preparing Water Management Master Plans at national and subnational levels in which natural conditions, technological solutions, nature-based solutions and social realities are considered.

第六节　科学支撑水治理

32　应当基于循证决策来增强社会韧性。促进有效水治理是一项重大挑战，是政府间水文计划第九阶段战略规划的基石，需要被大力支持，以便促使成员国实施循证决策，建设更具韧性、更繁荣的社区。同样，需要通过可靠的数据、有能力的人力资源和更多的伙伴关系，促进可持续水治理成为一项长期活动。因此，鼓励在基层发展以社区为基础的治理伙伴关系，能有效促进国家层面的政策转变。

33　新冠疫情全球健康危机引发了人们对可持续未来治理以及所需横向和纵向一体化的重新思考。该流行病为变革提供了机会，促进社会通过适应和转型“向前发展”。为了增强个人和社区韧性，全球倡导“以人为本”的新治理模式（国际应用系统分析研究所，2020 年）。其中一些模式适用于水务部门。管理和预防公共危机和水文气象灾害的经验表明，缺少全社会参与的政策和措施可能是无效的或容易被忽视的。因此，让民间团体和私营部门在可行的或需要的范围内参与治理机制是非常重要的。在这种情况下，应对水文气象灾害为在政府间水文计划第九阶段战略规划中拓展和加强水治理方案提供了一个机会，使其更具包容性、科学性，并让更广泛的利益相关方参与。

第七节　实现变化世界中的水安全

34　联合国教科文组织成员国将水安全定义为：保障流域内人口获得维持人类和生态系统健康、高质量、数量充足的水资源，并确保有效保护其生命和财产免受洪水、滑坡、地面沉降和干旱等涉水灾害影响。然而，广义上的“水安全”主要取决于如何分析水–社会挑战的复杂性。在复杂的自然和社会系统中，水安全的主要挑战在于未来水资源可用量和需水量的不确定性。

35　政府间水文计划第八阶段战略规划所做的贡献和产出的成果，是构建第九阶段战略规划新型结构内容的基础。同样，在过去八年中，政府间水文计划和联合国教科文组织水事机构建立了伙伴关系，为开展政府间水文计划第九阶段战略规划活动配备了一批经验丰富的实体和专业人员，他们多年来致力于促进和增强水安全。联合国教科文组织全体水事机构都为实现水安全作出了重大贡献。与此同时，在实施政府间水文计划第八阶段战略规划的过程中，分别在韩国和墨西哥设立了一个由联合国教科文组织赞助的二类水安全中心，在美国设立了一个联合国教科文组织可持续水安全问题教席。联合国教科文组织水事机构将继续促进和开展有关水安全的研究和能力建设活动，并为成员国提供专业知识方面的支持。

Scientific support to water governance

32 To enhance the resilience of societies, decisions should be evidence-based. Improving the effectiveness of water governance is a major challenge, which needs to be strongly supported as a cornerstone of IHP-IX to enable Member States to implement evidence-based decisions to build more resilient and prosperous communities. Similarly, there is a need to promote sustainable water governance as a long-term activity through sound data, capacitated human resources, and increased partnerships. Therefore, encouraging the development of community-based governance partnerships at the grassroots level can lead to effective policy changes at the national level.

33 The COVID-19 global health crisis has served as a trigger for re-thinking governance for sustainable futures and the horizontal and vertical integration required. The pandemic has opened a window of opportunities for change and for "bouncing forward" through adaptation and transformation. To enhance the resilience of individuals and communities, new people-driven governance models have been advocated (IIASA 2020). Some of these models can be adapted to the water sector. Managing and preventing public crises and hydrometeorological disasters show that without whole-of-society engagement policies and measures may turn out to be ineffective or ignored. Therefore, it is important that civil society and private sector be involved in governance mechanisms to the extent feasible or needed. In this context, coping with hydrometeorological disasters provides an opportunity to expand and enhance under IHP-IX, more inclusive, science-based water governance schemes involving a wider spectrum of stakeholders.

Attaining water security in a changing world

34 UNESCO's Member States defined water security as the capacity of a population to safeguard access to adequate quantities of water of an acceptable quality for sustaining human and ecosystem health on a watershed basis, and to ensure efficient protection of life and property against water-related hazards such as floods, landslides, land subsidence, and droughts. However, the definition of the full extent of 'water security' depends mostly on how the complexity of water-society challenges is analysed. The main challenge of water security resides in the uncertainty of water availability and demand in the future in the midst of complex natural and social systems.

35 The contributions and results obtained during the implementation of the IHP-VIII provide a unique opportunity for the IHP-IX to build on and mainstream these into the newly structured programmatic content. Similarly, the partnerships built over the past eight years by IHP and the UNESCO Water Family offer the opportunity for the activities of IHP-IX to be carried out by a group of experienced entities and professionals that have worked in promoting and advancing water security for various years. While the entire UNESCO Water Family has contributed significantly towards the achievement of water security, two Category 2 Water Security Centres under the auspices of UNESCO and a UNESCO Chair on Sustainable Water Security were established in the Republic of Korea, Mexico and the United States of America, respectively during the implementation of IHP-VIII. The UNESCO Water Family will continue to promote and carry out research and capacity building activities related to water security, and will extend the support of their specialized expertise to the Member States.

第二章 政府间水文计划第九阶段战略规划的制定背景和战略目标

BACKGROUND AND STRATEGIC OBJECTIVE OF IHP-IX

第一节　联合国教科文组织和政府间水文计划的优势

36　联合国教科文组织通过其五个部门和相关的科学计划来实现跨学科方法。它可以利用自然和社会科学、教育、文化、通讯、信息等互补领域的广泛专业知识和信息。此外，通过政府间水文计划和自 2000 年开始实施的世界水评估计划，联合国教科文组织在水方面积累了 50 多年的经验。

37　政府间水文计划专门致力于推进水研究与管理，以及对促进可持续和水资源综合管理至关重要的相关教育和能力建设工作。联合国教科文组织政府间水文计划提供了一个与水有关的科学和教育平台，使其他互补性网络倡议能够汇集研究机构、博物馆、工业发展设施、创新中心、科学家、成员国代表、决策者、政府官员、青年和其他人员分享知识并整合不同观点。政府间水文计划和联合国教科文组织水事机构包括了 169 个政府间水文计划国家委员会和联络人，联合国教科文组织水科学司及其所属的世界水评估计划、派驻各地办事处的区域水文学家、36 个二类中心以及超过 65 个涉水问题小组教席，为国际社会提供了水领域最全面的水科学家、管理者和从业人员。因此，最重要的是促进联合国教科文组织水事机构成员间的合作与伙伴关系，确保政府间水文计划国家委员会和联络人具备足够的能力和手段来实施计划。此外，至关重要的是设立涉水二类中心和教席，特别是在非洲这个被联合国教科文组织优先支持以及需要在科学、研究和创新方面加强人员能力的地区。

38　联合国教科文组织已经通过其水事机构与其他政府间和国际组织等全球和区域伙伴建立了不同层级的工作关系。此外，政府间水文计划还与私营部门和非政府组织伙伴建立了伙伴关系，目的是将科学研究和创新成果转化为实际应用并促进各级知识共享。

39　此外，在联合国水机制的协调下，联合国教科文组织政府间水文计划通过机构间关系与联合国系统内的其他组织和机构开展合作活动和计划倡议。在这些行动中，重要的是突出那些旨在加速落实《2030 年可持续发展议程》、实现可持续发展目标的行动。

第二节　确保变化环境下政府间水文计划战略的连续性

40　迄今为止，政府间水文计划已经制定并实施了八个阶段战略规划，每个阶段都是在前一阶段的基础上解决了成员国明确提出的全球性重要问题。这种演变体现了它从水文科学发展为综合科学体制上的成长，以便支持政策制定和社会发展。

41　政府间水文计划第八阶段战略规划虽然在科学及其决策过程中的应用方面已取得了实质性进展，但是该战略规划涉及的许多问题仍然没有得到解决，而它们是相互关联的。

42　政府间水文计划第九阶段战略规划将涉及与水安全和可持续水管理有关的、相互关联的五个优先领域。这样，水教育成为从第八阶段向第九阶段延续和过渡的主线，并且更加注重新技术与教育之间的关系。同样，摆脱以往单一学科方法束缚，政府间水文计划第九阶段战略规划将采用相互交织的跨学科方法，来解决第八阶段其他五个专题中未解决的问题。

43　从第八阶段到第九阶段的另一个极为重要的过渡要素将是 18 个旗舰倡议（附件 4），它们将通过其他倡议得到深化或互补，有助于提高水安全所需的水科学和能力。通过政府间水文计划水信息网络系统，

THE COMPARATLVE ADVANTA GE OF LINESCO AND ITS INTERGOVERN–MENTAL HYDROLOGICAL PROGRAMME

36 UNESCO's transdisciplinary approach is accomplished through its five sectors and related science programmes. UNESCO can leverage a wide range of expertise and information in complementary fields, such as the natural and social sciences, education, culture, as well as communication, and information. Furthermore, the Organization has accumulated more than 50 years of experience on water through the Intergovernmental Hydrological Programme and since 2000 through the World Water Assessment Programme (WWAP).

37 The Intergovernmental Hydrological Programme is devoted exclusively to advancing water research and management, and the related education and capacity development efforts considered essential to foster sustainable and integrated water resources management. UNESCO-IHP offers a scientific and education platform related to water, enabling other complementary network initiatives that bring together research institutes, museums, industry development facilities, innovation centres, scientists, Member States representatives, policy makers, government officials, youth and others, to share knowledge and integrate different points of view. IHP together with its UNESCO Water Family comprises the 169 IHP National Committees and focal persons, UNESCO's Division of Water Sciences, including the World Water Assessment Programme, regional hydrologists posted in field offices, its 36 Category 2 Centres, and more than 65 thematically grouped water-related UNESCO Chairs, offers the international community the most comprehensive grouping of water scientists, managers and practitioners in the water arena. It is thus of primary importance to facilitate the cooperation and partnerships of UNESCO Water Family members and ensure that IHP National Committees and focal persons have adequate capacity and means to contribute to the implementation of the Programme. Furthermore, it is crucial that the establishment of water-related Category 2 Centres and Chairs will be supported particularly in Africa, a UNESCO priority and a region where strengthening human capacity for science, research and innovation is needed.

38 Through this network, UNESCO has established working relationships with global and regional partners at various levels, including other intergovernmental and international organizations. Additionally, IHP has developed partnerships with the private sector and NGO partners aimed to advance the results of scientific research and innovation into practical uses and to promote knowledge sharing at all levels.

39 Furthermore, within the coordination mechanism of UN-Water and through inter-agency relations, UNESCO-IHP develops and undertakes collaborative activities and programmatic initiatives with other organizations and agencies of the United Nations system. Among these actions, it is important to highlight those aimed at accelerating the UN 2030 Agenda and the SDGs.

ENSURING CONTINUITY WITH CHANGE

40 To date, IHP has developed and implemented eight phases, each one building on the prior phase while addressing issues of global importance, as clearly expressed by Member States. This evolution represents an institutional growth from hydrological sciences to integrated sciences, supporting policy and society.

41 While substantive advances have been made in both science and its application to decision processes, many of the issues that were addressed in IHP-VIII are still unresolved and relevant.

42 IHP-IX will address five priority areas, all interconnected and related to water security and sustainable water management. In this manner, water education becomes a main axis of continuity and transition from phase VIII to phase IX, with increased importance on the relationship between new technologies and education. In the same way, unresolved issues from the other five themes of the eighth phase of the IHP are reflected in IHP-IX with an intertwined, transdisciplinary approach and steering away from the silo approach of the past.

43 An additional critically important transition element from IHP-VIII to IHP-IX will be the 18 Flagship Initiatives (see annex 4), which will be deepened or complemented with other initiatives, contributing to improve the water science and capacities required for water security. Through the IHP-Water Information Network System (IHP-WINS), efforts will be made to connect all IHP Flagship Initiatives data-related platforms as well as other relevant water data platforms.

可以连接所有政府间水文计划旗舰倡议数据平台以及其他相关的水数据平台。

44 同样，为了提高利益相关方的能力并增强其对极端水文气象灾害复原力，将继续在不同区域建立干旱观测站、在非洲开发和部署洪水预警系统。现有的气候风险知情决策分析方法将被用于识别世界各地新研究地点的水安全风险。此外，为了提供基于自然的城市发展替代方案，政府间水文计划第九阶段战略规划将继续支持在现有地点和新地点开展创新生态水文研究和应用。通过建立治理指南、多层次评估和科学合作，政府间水文计划第八阶段战略规划审议了地下水治理的关键方面，并特别注重与《水公约》协同的跨界地下水治理。在第九阶段，政府间水文计划将进一步开展活动，就地下水在支持弹性用水方面的重要作用进行研究和科学合作。它将继续协助成员国提高对地下水的科学认识、强化境内和跨界地下水治理框架。此外，第八阶段促进了对新兴污染物和微塑料的研究，并增强了相关知识库。为了应对这一新的全球水挑战，政府间水文计划将继续主导和推动新兴污染物和微塑料的研究和科学合作。虽然政府间水文计划第八阶段战略规划已对现有城市水系统、宏观城市水管理方法、水敏感城市设计和城市代谢开展了大量的最新审查，但是第九阶段将继续跟踪已找出的差距，同时推广智慧水管理系统，研究水在城市规划和循环经济中的作用。这些是政府间水文计划从第八阶段延续到第九阶段的一些旗舰倡议的例子。

45 基于前几个阶段的教训和第八阶段的经验，实施政府间水文计划第九阶段战略规划时将按国家和地区采取适应性方法，并在政府间水文计划水事机构各层级（理事会、政府间水文计划国家委员会、中心、教席、区域顾问单位等）之间开展强有力的全球协调。政府间水文计划项目将通过动态和适应性途径从第八阶段过渡到第九阶段，同时确保其连惯性。

第三节　政府间水文计划的战略目标

一、政府间水文计划的愿景

46 政府间水文计划的愿景是建立一个水安全的世界。在这个世界里，人民和机构有足够的能力和科学基础知识，能就水管理和水治理作出明智决策，以实现可持续发展，构筑韧性社会。

二、政府间水文计划第九阶段战略规划的使命

47 在 2022—2029 年期间，政府间水文计划的使命是支持成员国与水领域的合作伙伴和其他联合国机构开展积极合作，通过水科学和水教育加快落实涉水可持续发展目标和其他相关议程。

为此，政府间水文计划第九阶段战略规划将：

1）利用跨部门合作，实现一个水安全的世界。

2）促进国际科学研究与合作，融合人类与水系统之间相互作用，完善应对水挑战和气候变化所需的知识。

3）有效调动和传播与科学和政策相关的专业技能、知识和工具，以便在应对水挑战方面作出明智决策。

4）强化机构和人员能力，培训当代和下一代水利从业人员，为可持续发展目标提供水问题解决方案，并通过解决水问题来建立气候适应力。

5）提高各级意识，促进水文化和水道德建设，保护和节约水资源并促进可持续水资源综合管理。

6）考虑相互关联的水与气候问题，支持实现联合国可持续发展目标 6，落实“全球加速器框架”。

7）通过对利益相关方参与、跨学科知识整合的方法研究的支持，加强跨学科的水研究。

44 Similarly, the drought observatories established in different regions and the flood early warning systems developed and deployed in Africa will be continued to improve stakeholder capacity and enhance resilience to extreme hydro meteorological phenomena. The developed Climate Risk Informed Decision Analysis (CRIDA) methodology will be applied in new study sites around the world for identifying water security risks. In addition, IHP's support of innovative ecohydrological research and applications in present and new sites will continue during the ninth phase to provide nature-based urban development alternatives. The IHP-VIII phase considered key aspects of groundwater governance, with a particular focus on transboundary in collaboration with the Water Conventions, by establishing governance guidance, assessment at multiple levels, and scientific cooperation. During phase IX, IHP will further develop activities dedicated to research and scientific cooperation on the essential role of groundwater to support resilient water use. It will further continue to assist Member States in improving the scientific knowledge on groundwater and in strengthening groundwater governance frameworks at domestic and transboundary level. Furthermore, the IHP-VIII phase has promoted research and enhanced the knowledge base on emerging pollutants and microplastics. IHP will continue to play a leading role in promoting further research and scientific cooperation on emerging pollutants and microplastics in order to respond to this new global water challenge. Whereas a number of state-of-the-art reviews of existing urban water systems, approaches to macro urban water management, water-sensitive urban design and urban metabolism have been conducted during IHP-VIII, gaps identified will be pursued in IHP-IX, while the promotion of Smart Water Management Systems, researching the role of water in urban planning and circular economy will continue. These are examples of some of the flagships that will continue from phase VIII of IHP into phase IX.

45 Building on the lessons learned from previous stages and the experience gained from the IHP-VIII phase, the implementation of IHP-IX will adopt an adaptive approach by country and region and a strong global coordination among the IHP Water Family at all levels (Council, IHP National Committees, Centres, Chairs, regional consultation units, etc.). The IHP programmatic undertakings will transition from IHP-VIII to IHP-IX, via a dynamic and adaptive pathway, while ensuring continuity.

IHP VISION

46 IHP envisions a water secure world where people and institutions have adequate capacity and scientifically based knowledge for informed decision-making on water management and governance to attain sustainable development and to build resilient societies.

IHP–IX MISSION

47 Our mission for the period 2022—2029 is to support the Member States to accelerate the implementation of water-related SDGs and other relevant agendas through water science and education in cooperation with partners and other UN agencies active in the water sector.

To this end, IHP-IX will, in its ninth phase:

a. Leverage intersectorality for a water secure world.

b. Promote international scientific research and cooperation for improved knowledge to address water challenges and climate changes incorporating the interaction between human and water systems.

c. Mobilize and disseminate effectively scientific and policy relevant expertise, knowledge and tools for informed decisions in addressing water challenges.

d. Reinforce institutional and human capacities and train the present and upcoming generation of water professionals capable of providing water solutions for SDGs and building climate resilience through water.

e. Raise awareness and promote a water culture and water ethics at all levels for the protection and conservation of water resources and the promotion of sustainable integrated water resources.

f. Support the achievement of UN SDG 6 and the implementation of the Global Acceleration Framework, taking into account interrelated water and climate issues.

g. Strengthen transdisciplinary water research, by supporting research on methods for stakeholder involvement and transdisciplinary knowledge integration.

第三章
政府间水文计划
第九阶段战略规划
预期成果和优先领域

OUTCOME AND PRIORITY AREAS OF IHP-IX

48 实现水安全带来了若干挑战，这些挑战包括全球变化影响（如水灾害）和业务方面的问题（例如通过当地水价理解水的价值）。政府间水文计划应对这些挑战的方法是，拓展各级人员潜力、科学基础与知识,以“理解全球变化对水系统的影响，并将科学结论用于制定促进可持续水资源管理的政策”。

49 总体预期成果：政府间水文计划支持成员国“基于改进的科学数据、研究、知识、能力和科学－政策－社会交互关系，践行包容和以循证水治理与管理，建设可持续、韧性社会”。

50 上述第九阶段预期成果与联合国教科文组织总体中期战略（41 C/4，2022—2029 年）保持一致，并将服务于它的两个战略目标：

51 战略目标 1：确保优质、公平和包容的教育，促进人人享有终身学习机会，减少不平等现象，特别是在数字时代促进学习型和创造型社会建设；战略目标 2：通过促进科学、技术、创新以及保护自然遗产，努力建设可持续社会，保护环境。

52 循证水治理和管理的一个先决条件是具备见多识广、训练有素、有能力的专业人员提供的可获取的、可利用的、最新的科学知识。

53 一方面，提升科学家、决策者和从业人员之间的合作水平，促进公民对科学活动的贡献（这里的公民科学是指收集数据和定义待解决问题的科学）对弥合现有数据与政策领域内人员之间的鸿沟至关重要。那些人员必须首先理解、随后解释和应用这些技术信息。然而，为了实现这种有助于进行更好的循证水管理的合作，必须验证数据收集和分析方法，并投入充足的时间和精力，在不同的地理和政治背景下开发不同尺度的综合数据。

54 另一方面，需要开展当代人的能力建设，培养下一代各级水规划者、科学家、决策者和从业人员，以及对水敏感的公众。通过开展水教育和拓展水知识来提升能力建设，提高公众对可持续水文化的认识，使其为可持续生产和消费改变行为、建立共识，最终使经济增长与环境退化脱钩。此外，基于开放型软件平台的开源决策支持系统在水资源管理中发挥了日益重要的作用。

55 改善水管理决策和水治理，需要通过完善科学、推动合作来弥合水数据与知识之间的鸿沟。毫无疑问，所生成知识的质量将直接反映在所制定政策的可持续性上。尤其在全球变化背景下，经得起时间考验的水政策证明了社会具有解决当今社会复杂水问题所需的韧性。

56 应对全球变化的战略和活动应以科学为基础，囊括社会所有部门，以便全面提升社会韧性。面对不断变化且更加复杂的环境条件，需要利用科学政策来建设韧性社区和社会。为了完善决策过程，需要公民科学和积极主动的非政府组织、民间团体和社区伙伴的参与，包括利用土著知识提高决策者能力。

57 为了监督预期成果进展，本战略规划确定了下列业绩指标：

业绩指标 1：改善水科学和研究，拓展知识、提升各级服务和相关风险管理水平的成员国或利益相关方的数目。
业绩指标 2：加强各级非正式、正规和非正规水教育的成员国数目。
业绩指标 3：使用、开发并鼓励运用科学的、质量可控的数据和知识实现可持续地水资源管理的成员国数目。
业绩指标 4：为应对全球挑战而实施水资源综合管理的成员国数目和实施程度。
业绩指标 5：为了加强缓解性、适应性和韧性水治理，科学实施相关机制、政策和工具的成员国数目和实施程度。

48 Several challenges arise from achieving water security, which range from the effects of global change such as water-related disasters to operational aspects such as understanding the value of water as this is expressed by local water rates. The Intergovernmental Hydrological Programme's approach to these challenges is to expand the human potential, scientific base and knowledge at all levels to "understand the impacts of global changes on water systems and to link scientific conclusions to policies for promoting sustainable management of water resources".

49 Overall outcome: IHP will do so by providing support to its Member States to "practice inclusire and evidence-based water governance and management based on improved scientific data, research, knowledge, capacities and science-policy-society interfaces towards sustainable resilient societies".

50 The above identified outcome of IHP-IX is aligned with UNESCO's overall Medium-Term Strategy, 41 C/4 (2022—2029) and will serve two of its Strategic Objectives:

51 Strategic Objective 1: Ensure quality, equitable and inclusive education and promote lifelong learning opportunities for all, in order, inter alia, to reduce inequalities and promote learning and creative societies, particularly in the digital era and Strategic Objective 2: Work towards sustainable societies and protecting the environment through the promotion of science, technology, innovation and the natural heritage.

52 A prerequisite to evidence-based water governance and management is available, accessible and current scientific knowledge provided by informed, trained and capacitated human resources.

53 Enhancing the level of cooperation among scientists, policy-makers and practitioners, and the contributions of citizens to scientific activities (to data collection and the definition of problems to be solved——citizen science) is vital to bridging the gap between existing data and those who must understand and subsequently interpret and apply this technical information in the policy arena. However, for this cooperation to become a reality, data collection and analysis methodologies must be validated and sufficient time and effort must be invested to develop comprehensive data at different scales and in distinct geographic and political settings, all contributing to better evidence-based water management.

54 Developing capacity for the current generation and educating the next generation of water planners, scientists, policy makers and practitioners along with a water sensitized public, at all levels of sophistication is the other half of the equation. Water education and the expansion of knowledge about water make it possible to improve capacity development and expand public awareness towards a sustainable water culture, change behaviours and build consensus for sustainable production and consumption that results from decoupling economic growth from environmental degradation. In addition, open source decision support systems built on open software platforms play an increasingly important role in managing water resources.

55 Bridging the water data and knowledge gaps through improved science and cooperation will result in improved water management decisions and governance. The quality of knowledge generated is without a doubt directly reflected in the sustainability of policies developed. Water policies that stand the test of time, particularly in the context of global change, is an example of the type of societal resilience required to address the complex water issues facing society today.

56 Strategies and activities addressing global changes that are science-based and inclusive of all sectors of society, enhance the overall resilience of these societies. Building communities and societies that are resilient in the face of changing and evermore complex environmental conditions, requires that science inform policy. Improving this aspect of the decision process permits involvement of citizen science and pro-active NGO, civil society and community partners with government, including the ability of decision makers to benefit from the use of indigenous knowledge.

57 The following Performance Indicators (PI) have been identified to monitor progress towards the achievement of the desired Outcome:

PI 1: Number of Member States/stakeholders use improved water science, research and apply the strengthened capacities to expand knowledge and better manage services and related risks at all levels.

PI 2: Number of Member States with enhanced water informal, formal and non-formal education at all levels.

PI 3: Number of Member States which use, develop and encourage scientific and quality-controlled data and knowledge to sustainably manage their water resources.

PI 4: Degree of integrated water resources management addressing global challenges practice by number of Member States.

58 本战略规划是在联合国教科文组织秘书处的支持下由成员国和水事机构成员起草和拓展的。根据第八阶段战略规划的中期评估建议，联合国教科文组织水事机构成员（36 个二类中心、60 多个学术中心涉水教席以及 169 个政府间水文计划国家委员会）承担实施第九阶段战略规划的职责。

59 政府间水文计划第九阶段战略规划的根本任务是持续发展政府间水文计划国家委员会，使其成为公共水事机构、学术和科学中心以及日益增多的非政府水组织的议事场所。

60 为了彰显其促进政府间水文计划第九阶段战略规划实施的强大能力，水事机构必须加强与成员国之间的讨论和共同提案，同时，在执行其他国际协议文书（如《2030 年可持续发展议程》《仙台框架》《巴黎协定》《新城市议程》和其他相关协议）时也要这样做。因此，第六项业绩指标专门用于监督政府间水文计划国家委员会在实现国家、区域和全球层面预期成果中的贡献：

业绩指标 6：引领国家、区域和全球层面水议程的水事机构成员的数目。

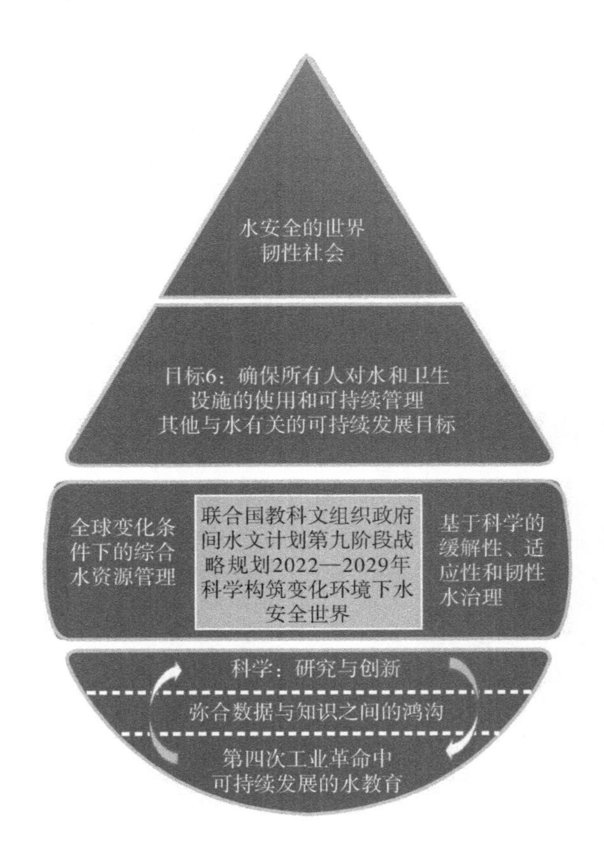

第一节　成果链与变革理论

61 实现水安全正逐渐被人们所理解，并成为全球关注的问题，其主要原因是水资源匮乏程度加重、水质日益恶化、洪涝和干旱极端天气事件的强度和频率不断增加，以及全球变化对人类生计、健康和环境的影响，和水资源对人类和平与安全的潜在影响。人类活动与经济增长共同增加了水资源和供水的压力，构成了水安全挑战的主要驱动力，并常常造成环境破坏。

PI 5: Degree of mechanisms, policies and tools based on science implementation to strengthen water governance for mitigation, adaptation and resilience by number of Member States.

58 The Strategic Plan presented herein originated from and was developed by Member States and the UNESCO Water Family Members with the support of the Secretariat. The mid-term evaluation of IHP-VIII recommends that during the execution of IHP-IX, roles and responsibilities for the implementation of the Plan are taken up by the UNESCO Water Family Members (36 Category 2 Centres, more than 60 water-related UNESCO Chairs linked to academic centres, and 169 IHP National Committees).

59 It is essential that IHP National Committees continue to develop so that their bodies act as meeting spaces between public water bodies, academic and scientific centres and increasingly, non-governmental water organizations.

60 To demonstrate a significant capacity to contribute to the implementation of IHP-IX, it is necessary for the Water Family to strengthen its insertion and contribution in the debate and proposals with the Member States and, simultaneously, to do so in other internationally agreed instruments such as the 2030 Agenda, the Sendai Framework, the Paris Agreement, the New Urban Agenda, and other relevant instruments. Consequently, a sixth Performance Indicator is dedicated to monitoring the contribution of IHP National Committees towards achieving the proposed Outcome via their role at multiple levels:

PI 6: Number of Water Family members leading the water agenda at national, regional and global levels.

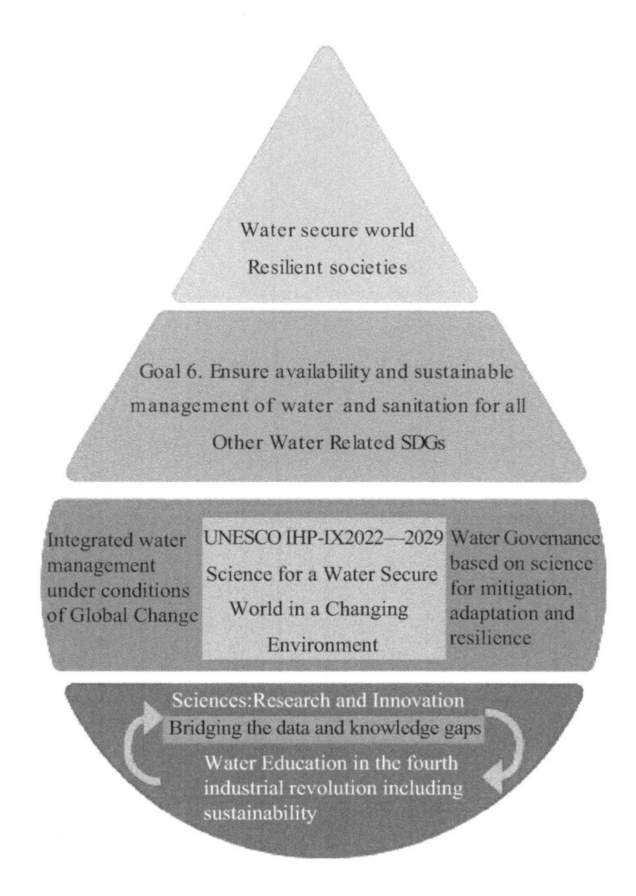

RESULTS CHAIN THEORY OF CHANGE

61 Achieving water security is gradually being understood and is becoming a global concern because of increasing water scarcity, degrading water quality, increased intensity and occurrence of extreme weather events such as floods and droughts, and the effects global changes have on human livelihoods, health,

62 政府间水文计划第九阶段确定了一项预期成果，如第 49 段所述：

1. 基于改进的科学数据、研究、知识、能力、适当工具和科学—政策—社会交互关系，（成员国）践行包容和循证水治理与管理，建设可持续的、韧性社会。

63 预计这一成果将有助于实现联合国教科文组织的战略目标和政府间水文计划第九阶段战略规划的愿景。实现这一成果的关键要素是下文将详细介绍的五个优先领域的一连串行动。联合国教科文组织成员国把相互联系、互为补充的三个优先领域，即科学、研究与创新（优先领域 1）、第四次工业革命中可持续发展水教育（优先领域 2）以及弥合数据与知识之间的鸿沟（优先领域 3），作为水资源综合管理（优先领域 4）和基于科学的缓解性、适应性和韧性水治理（优先领域 5）的基础。

64 联合国教科文组织水事机构及其伙伴将制定包括财务战略在内的业务实施计划，用于监督战略规划的执行进度。业务实施计划将详细介绍每个优先领域中包含的具体成果及相关活动。

65 计划开展大量的与研究、评估、基准创建有关的活动，通过共同商定协议收集有效数据，并通过使用共同商定的方法来分析这些数据，从而凝练成知识。利用第四次工业革命开发的新知识和技术，加强人力资源的各层次教育（中等教育、职业教育、高等教育和终身教育）和各类教育（非正式、正规和非正规）活动。这些训练有素的科学家、教师以及其他知识经纪人、知情公众和决策者将成为新的水文化变革的推动者，而这种新的水文化将支持可持续发展。这些活动旨在为管理和政策决定提供信息，并根据全球变化影响的复杂需求来改革机构设置。

66 这些活动将响应各成员国的诉求，需要人力和财力资源，以及与科学界、其他联合国机构、科学和专业协会、非政府组织和公众的合作与协作。开展这些活动的先决条件是理解和接受新的知识工具和产品。其次，接触直接受益方，采用定制的、合适的物质资料将助力变革所需的知识转移。再次，公众和决策者必须愿意利用所提供的知识和支持，通过开展活动将其转化为成果。最后，为了实现预期结果，资金支持、科学界和其他伙伴支持跨学科研究的意愿也非常重要。

67 所有这些努力都将有助于建立一个：“水安全世界。即为了实现可持续发展和建设韧性社会，人类和机构有足够能力和科学基础知识就水管理和治理开展明智决策”（参见第 46 段的政府间水文计划愿景）。

第二节　优先领域

68 联合国教科文组织成员国确立和阐述的政府间水文计划第九阶段战略规划优先领域以下列五个变革工具的形式呈现。这些工具将在 2022 年—2029 年期间，在不断变化的世界中，促进水安全，实现可持续发展：

1）科研与创新

2）第四次工业革命中可持续发展水教育

3）弥合数据与知识之间的鸿沟

4）全球变化条件下的水资源综合管理

5）基于科学的缓解性、适应性和韧性水治理

environment, and potential impacts on peace and security. The primary driver for these challenges is human activity, which, along with economic growth, has increased the pressures on water resources and supply and quite often to the detriment of the environment.

62 The ninth phase of IHP has one identified Outcome, as noted above (Paragraph 49):

1. Member States practice inclusive and evidence-based water governance and management based on improved scientific data, research, knowledge, capacities, appropriate tools and science-policy-society interfaces towards sustainable resilient societies.

63 It is expected that this Outcome will contribute to the achievement of UNESCO's Strategic Objectives, and to achieve IHP-IX's Vision. Five priorities, streams of action have been identified as the key elements to materialize this Outcome and are presented in detail hereafter. UNESCO's Member States identified as Priority Areas Science, Research and Innovation (Priority 1), Water Education for the Fourth Industrial Revolution including Sustainability (Priority 2), and Bridging the data and knowledge gaps (Priority 3), as intertwined elements that feed each other and formulate the basis for Integrated Water Resources Management (Priority 4) and Water Governance based on science for mitigation, adaptation and resilience (Priority 5).

64 Each Priority consists of several identified Outputs that along with activities, will be further developed within an operational document together with a financing strategy to be elaborated at a later stage by UNESCO's Water Family and its partners. The document will be used to monitor the implementation progress of the Strategic Plan.

65 Numerous activities are planned related to research, assessments, creation of baselines to generate knowledge based on validated data collected via commonly agreed protocols and analysed via commonly agreed and used methodologies and approaches. Activities related to strengthening the human capital at all education levels (secondary, vocational, tertiary, life-long) and for all education types (formal, informal, non-formal), will benefit from the new knowledge developed and from the technological opportunities offered by the fourth industrial revolution. The trained scientists, teachers, and other knowledge brokers, the informed public and decision-makers will become agents of change towards a new water culture that supports sustainable development. These activities aim at informing management and policy decisions, and reform institutional settings to align them with the needs as they become more complex due to the effects of global change.

66 The activities will respond to expressed Members States' needs and require human and financial resources, cooperation and collaboration with the scientific community, other UN Agencies, scientific and professional associations, NGOs and the general public. Understanding and uptaking of the new knowledge tools and products produced is a prerequisite. Reaching the direct beneficiaries and use of customized, appropriate materials will facilitate the knowledge transfer required for change. The willingness of the public and decision-makers to use the knowledge and support provided will be necessary in implementing activities and transform them into results. Similarly, the availability of finances and the willingness and support of the scientific community and other partners producing transdisciplinary research are invaluable in achieving the desired result.

67 The impact of all efforts will be to achieve "a water secure world where people and institutions have adequate capacity and scientifically based knowledge for informed decisions on water management and governance to attain sustainable development and to build resilient societies" (IHP Vision, see Paragraph 46).

PRIORITY AREAS

68 The IHP-IX Priority Areas, identified and elaborated by UNESCO's Member States, are presented as five transformative tools that will enable water security to sustain development in a changing world for the period 2022—2029:

1. Scientific research and innovation
2. Water education in the Fourth Industrial Revolution including Sustainability
3. Bridging the data-knowledge gap
4. Integrated water management under conditions of global change

69　为了研提和实施这五个优先领域中的每一领域并取得预期成果，不仅需要聚焦每个主题，提升和增加可持续水管理的价值，而且需要通过不同领域间的相互联系和协同作用来实现这一预期成果：“基于改进的科学数据、研究、知识、能力、适当工具和科学—政策—社会交互关系，成员国践行包容和以循证水治理与管理，建设可持续的韧性社会”。此外，深化和实施任一优先领域都将有助于落实《2030年可持续发展议程》及实现其17项可持续发展目标，因为这些目标遵循的原则都是：全面、平衡、可持续、公平、普遍和不可分割。由于政府间水文计划第九阶段战略规划结束年份（2029年）恰逢《2030年可持续发展议程》收尾，因此，在这个十年结束前，必须全面实现这些优先领域目标，并将相应成果转化为促进经济增长、社会包容和环境保护的三个可持续发展因素。

一、优先领域1：科研与创新

70　水文科研的发展为社会提供了有关水通量、输运和管理的实用知识和信息，然而，日益增长的不确定的环境变化要求我们在研究创新和应用方面继续努力。面对复杂的水科学和管理问题，科学研究考虑了人类与自然的相互作用，与新工具、新方法和新技术的应用共同为水资源管理提供了基本反馈。

71　到2029年，成员国将拥有知识、健全的科学和研究能力、新的和改进的技术以及管理技能，以确保人类发展和生态系统健康，实现可持续发展的水资源安全。

优先领域1与《2030年可持续发展议程》之间的关系

72　《2019年全球可持续发展报告》明确界定了优先领域“科研与创新”与可持续发展目标之间的关系。这份报告由联合国秘书长指定的独立科学家小组起草并发表在高级别政治论坛“现在就是未来：科学实现可持续发展目标”上。该报告强调，应对气候变化（可持续发展目标13：气候行动）、降低在获取维持生命的资源方面的不平等性（可持续发展目标6：清洁水和卫生设施，可持续发展目标7：负担得起的清洁能源和可持续发展目标9：工业、创新和基础设施）以及全面实现可持续发展目标离不开科技创新。此外，可持续发展目标12（可持续消费和生产）对于减少污染及其对水的影响以及提高用水效率至关重要。如果缺少基础科学和对问题的透彻理解，特别是对水问题的透彻理解，就无法解决贫困（可持续发展目标1）或饥饿（可持续发展目标2）问题。

预期成果

1.1　加强和促进国际科学合作[3]，提升对跨流域和含水层水文循环的科学认知以解决水文方面尚未解决的问题。

73　与水有关的社会问题变得越来越复杂，精简科学界议程也越来越受到国际科学界的重视。为了协调科学机构、专业组织和水利专业人员解决未解决的水文学问题，政府间水文计划在搭建科学平台和促进科学合作方面的作用日益重要。解决未解决的水文学问题，需要从根本上理解人与水之间的关系，以及水文过程与水文情景的协同演变关系。

74　其中一个未解决的水文学问题是水文过程的尺度和异质性研究，科学家和管理者为此激烈争论了几十

3　Blöschl，G. 等人（2019年），水文学的二十三个未解决问题(UPH)–水文界观点，《水文科学杂志》，第64卷第10期，第1141–1158页。https://doi.org/10.1080/02626667.2019.1620507

5. Water governance based on science for mitigation, adaptation, and resilience

69 Developing and implementing each of these five Priority Areas with their expected outputs implies advancing and adding value to sustainable water management not only from each of these thematic axes, but also through their interlinkages and synergies to achieve the expected Outcome: "Member States practice inclusive and evidence-based water governance and management based on improved scientific data, research, knowledge, capacities (...) and science-policy-society interfaces towards sustainable resilient societies". In addition, the deepening and implementation of each of the Priority Areas contributes to achieve the 2030 Agenda and its 17 SDGs, since all of them are governed by the principles of comprehensiveness, balance, sustainability, equity, universality and indivisibility. Since the conclusion of IHP-IX (2029) coincides with the end of the 2030 Agenda, it is essential that the contributions of these Priority Areas be fully implemented and translated into improvements in the three dimensions of sustainable development: economic growth, social inclusion and environmental protection, prior to the end of the decade.

PRIORITY AREA 1 : SCIENTIFIC RESEARCH AND INNOVATION

70 The development of hydrological science and research has provided practical knowledge and information for society about water fluxes, transport and management, however ever increasing and uncertain environmental changes demands for a continued effort on research innovation and application. Scientific research incorporating human interactions with nature in the context of complex water sciences and management problems provide fundamental feedback for water resources management, along with the application of new tools, approaches and technologies.

71 By 2029, the Member States have the knowledge, sound scientific and research capacity, new and improved technologies, and the management skills that allow them to secure water resources for human development and healthy of ecosystems within a sustainable development context.

Relationship between Priority Area 1 and the Agenda 2030

72 The link between Priority Area "Scientific research and innovation" and the SDGs was clearly defined in the 2019 Global Sustainable Development Report produced by an independent group of scientists appointed by the United Nations Secretary-General informing the High-Level Political Forum (HLPF) "The Future is Now: Science for Achieving Sustainable Development". This document stresses that scientific innovation is indispensable for addressing climate change (SDG 13 Climate Action), reducing inequalities in access to the resources that sustain life (SDG 6 Clean Water and Sanitation, SDG 7 Affordable and Clean Energy, and SDG 9 Industry, Innovation and Infrastructure) and achieving the SDGs in general. Moreover, SDG 12 (Sustainable Consumption and Production) is paramount in reducing pollution and its impacts on water and in enhancing efficiency in the use of water. You cannot solve poverty (SDG 1) or hunger (SDG 2) without the underlying science and a thorough understanding of the problem and specifically in relation to water.

Expected outputs:

1.1. International scientific cooperation strengthened and fostered to address unsolved problems in hydrology[3], improving scientific understanding of hydrological cycles across river basins and aquifers.

73 Since societal problems related to water have become ever more complex, streamlining a science community agenda is more important than ever as recognized by the international scientific community. IHP's role is increasingly important to facilitate a scientific platform and foster scientific cooperation to address Unsolved Problems in Hydrology (UPH) in coordination with scientific institutions, professional organization and water professionals. Addressing UPH requires fundamental understanding of the human water nexus and coevolution of hydrologic processes and scenarios.

74 One of the UPH is scaling and heterogeneity research in hydrological processes that has engendered a spirited

---

3 Bloeschl el al. Twenty-three unsolved problems in hydrology (UPH) – a community perspective. HYDROLOGICAL SCIENCES JOURNAL 2019, VOL. 64, NO. 10, 1141–1158. https://doi.org/10.1080/02626667.2019.1620507

年。科学家们仍在努力研究各种物理和生物因素与水文变量和通量的空间均质性和异质性之间的关系。如何在点、山坡、流域和大陆等不同尺度上应用水文学原理，并在不同时空尺度上关联这些数据，仍然是一个悬而未决的问题。政府间水文计划第九阶段战略规划将与科学机构、国际水文科学协会、国际水文地质协会、国际水利与环境工程学会、世界泥沙研究协会等专业科学组织，以及世界气象组织等有关联合国机构合作，为解决未解决的水文学问题提供科学平台和合作支持。

75 政府间水文计划第九阶段战略规划将强调使用新型监测技术，特别是最新的信息和通信技术、遥感技术和大数据，这些技术为观测和模拟大范围时空尺度上的水文过程提供令人振奋的机遇。

76 此外，政府间水文计划将动员国际科学界推动水文学研究，包括对水文学与其他学科相互关系的研究，以便激发科学创新活力，在地方、区域和全球层面解决与水资源有关的问题。

1.2 在联合国教科文组织指定地点开展生态水文学研究和创新工作，实现科学界和联合国教科文组织水事机构成果共享和交流，评估生态水文学和基于自然的解决方案对水循环的影响，并将此类方案纳入水资源综合管理、所有规模的服务和指定地点的管理之中。

77 基于自然的生态水文学解决方案有助于改善水管理、保护生态敏感的环境并提供关键性的服务，例如用于处理污水和减缓洪水影响的湿地，以及用于减少海浪、风暴潮和海岸侵蚀影响的红树林。生态水文学原理提供了一个利用生态系统过程来管理流域的框架，但仍未能解答许多研究问题。应用以下生态水文学三个主要原则，可更好地理解这些问题的复杂性：1）量化水文学和生物学过程，描述威胁的特征；2）利用生态系统属性和服务作为水管理工具（生态水文学和基于自然的解决方案）；3）协调灰色和绿色基础设施，实现与水资源综合管理密切相关的生态系统可持续性。

78 自政府间水文计划第六阶段战略规划以来，支持生态水文学研究一直是联合国教科文组织的优先事项，第九阶段战略规划将继续这一优先事项。利用越来越多的联合国教科文组织指定地点，政府间水文计划第九阶段战略规划将支持开展生态水文学研究，分享和交流研究成果，为各种尺度的水资源综合管理和服务提供解决方案。

1.3 开展气候情景、水文预测和用水情景的不确定性研究，详细制定供决策者和公众参考的适应性水管理策略。

79 正确理解水文气候不确定性的来源和影响，重新认识历史趋势的平稳性，为水资源管理和规划决策提供可靠的水文预测依据。此外，最新数据表明，持续不断的社会经济和非环境因素不确定性也将影响水资源系统。需要对气候和非气候因素不确定性进行分类，让利益相关方参与风险评估，以便制定适应性水管理战略。

80 基于以往涉水风险不确定性的工作，政府间水文计划第九阶段战略规划将继续支持政府机构和流域管理机构应用不确定性分析和涉水风险概率映射的最新进展。此外，水文预测、风险沟通以及利益相关方的不确定性，将会被视为管理涉水风险战略不可或缺的组成部分。

81 政府间水文计划将促进各种方法的应用，以便评估气候和非气候因素不确定性对水资源的影响，以便在对以往知识不足以预测未来时，制定有效的、稳健的能应对上述不确定性的战略规划。

82 此外，政府间水文计划第九阶段战略规划还将强调几种韧性水管理工具和方法的应用，这些工具和方

debate between scientists and managers for decades. Scientists still struggle with the relationship between various physical and biological factors and spatial homogeneity and heterogeneity in hydrological variables and fluxes. It is also an open question how hydrological principles should be applied at different scales (e.g., point-scale, hill-slope-scale, catchment-scale, and continental-scale) and how to relate such data when scales change over space and time. IHP-IX will facilitate provide scientific platform and support cooperation to address UPH, in partnership with scientific institutions, professional scientific organizations such as International Association of Hydrological Sciences (IAHS), International Association of Hydrogeologists (IAH), International Association for Hydro-Environment Engineering and Research (IAHR), WASER and relevant UN organizations, such as the World Meteorological Organization.

75 IHP-IX will emphasize the use of new monitoring techniques, and in particular, the latest ICT technologies, remote sensing and big data, which offer exciting opportunities for observing and modelling hydrological processes across a wide range of spatial and temporal scales.

76 Furthermore, IHP will mobilize the international scientific community to advance hydrological research including working on interfaces between hydrology and other disciplines to stimulate scientific and innovative undertakings required to address questions related to water resources at local, regional and global level.

1.2. Ecohydrology research and innovation at UNESCO-designated sites conducted and shared by the scientific community and UNESCO Water Family, communicated to assess the impact of ecohydrological and nature-based solutions on water cycles and include such solutions in Integrated Water Resource Management (IWRM) and services at all scales and in Sites' management.

77 Nature-Based Solutions (ecohydrology) contribute to the improved management of water and enabling protection of ecologically sensitive environments, providing critical services, such as wetlands for sewage treatment and flood mitigation, or mangroves to reduce the impact of waves, storm surge and coastal erosion. The principles of ecohydrology provide a framework as to the use of ecosystem processes as basin management tools, but many research questions remain unanswered. Greater understanding of these complexities will be achieved by applying the three main principles of ecohydrology: the quantification of both hydrological and biological processes, the characterisation of threats, use ecosystem properties and services as a tool in water management (Ecohydrology and Nature-based Solutions) and the harmonization of grey and green infrastructures to achieve sustainability of ecosystems closely related with IWRM.

78 Supporting research in ecohydrology has been a priority for UNESCO since IHP-VI and will continue being so in IHP-IX. Using an expanding number of UNESCO-designated sites, IHP-IX will support conduct of ecohydrology research, share and communicate its results to provide solutions in Integrated Water Resource Management (IWRM) and services at all scales.

1.3. Research on uncertainty in climatic scenarios, hydrological projections and water use scenarios conducted and recommendations communicated to decision makers and the general public to elaborate adaptive water management strategies.

79 A proper understanding of the sources and effects of hydro-climatic uncertainty, with the reconsideration of stationarity of historical trend, is required to develop reliable hydrological projection as the basis for decision making in water resource management and planning. Furthermore, the recent past has demonstrated that, societies are also constantly facing socioeconomic and non-environmental uncertainties which will impact water resources systems. Both climatic and non-climatic uncertainties need to be classified and addressed through risk assessment including stakeholder engagement to develop adaptive water management strategies.

80 Based on the previous work on uncertainties of water-related risk, IHP-IX will continue to support application of recent advances in uncertainty analysis, and probabilistic mapping of water-related risks among government agencies, and river basin authorities. Moreover, uncertainty in hydrological projection and risk communication and stakeholder participation will be further promoted as an integral part of strategies for managing water-related risks.

81 IHP will promote application of methodologies to assess the impact of climatic and non-climatic uncertainty on water resources and to work towards effective adaptation strategies for planning for robustness and adaptability under uncertainty as such, when knowledge about the past is not sufficient to predict the future.

82 Furthermore, IHP-IX will also emphasize the application of several resilient water management tools and

法可以评估集成在水文预测和用水情景中的水文气候和非气候因素的不确定性，以便支撑水资源规划和管理决策支持系统。

1.4 支持科学界就新商业模式、水务公司作用、广泛参与和伙伴关系、基础设施等方面开展科学研究和探索，加速水务部门向循环经济转型。

83 当今水务部门和大多数流域仍然普遍采用线性水管理模式（取—用—排），这构成了全世界水资源不可持续生产和消费的基础。循环经济是水资源综合管理的重要组成部分，也考虑了城乡联系以及城市生物圈内的水资源系统。

84 向循环经济转型应考虑整个价值链的资源消耗和生产，以便在水循环内创造协同效应，实现更公正、更有效的水资源综合管理。使用循环代谢方法可进一步促进不同尺度的更加系统化的水资源综合管理，特别是通过运用基于自然的解决方案和适当考虑综合生态系统服务，更好地使人类水循环与自然水循环相一致，从而达到加强和保护资源的目的，特别是缓解和适应气候挑战的目的。

85 政府间水文计划第九阶段战略规划将通过研究若干有利因素来支持向循环经济转型，这些因素包括：面向系统的科学和技术进步、新的商业模式和水务公司的作用、有利于公正转型的广泛参与和伙伴关系，以及对现有基础设施改造设计和新的基础设施设计的重新思考等等，优化用水并产生协同效益，例如有关水－能源－粮食－生态系统关系的研究。现有基础设施的改造设计，需要充分发挥资源效率和回收能力、优化节能减排，实现跨部门共享基础设施。

1.5 支持科学界根据社会水文学对人与水之间的相互作用进行评估并分享评估成果，制定水管理的适应性途径、管理方案和战略。

86 由于人类活动对水资源的影响，水与人类的关系变得非常紧密。持续的人类与复杂水管理问题的相互作用引发了许多新问题和可能性，这些是水文科学家不能独自解决的。

87 社会科学在有效运用“协同创新”和“协同设计”的技术和方法方面发挥着越来越重要的作用，并被证明是在欠发达地区和较小村庄引进新技术的有效方式。

88 社会水文学提供了人类与水系统之间的双向反馈，由此导致了世界不同地方和不同环境下出现的各种现象。

89 支持社会科学和自然科学的更大重合，为项目的设计和实施创造了更好的条件，以便解决在全球变化及其社会影响条件下水文循环变异性和变化等复杂问题，它为应对更具破坏性的水文灾害、更好地管理水－能源－粮食－生态系统关系、城市－区域水代谢以及水资源短缺和水资源系统，包括跨界水资源系统，提供了一个更好的适应性决策基础。

90 通过加强理解水与社会之间相互作用的动态性，政府间水文计划第九阶段战略规划将基于支持公平解决方案的科学成果，促进解决涉水社会问题，实现水安全。它还将支持识别与水管理相关的社会目标之间的协同和平衡效应。政府间水文计划将在成员国内建设产出这类研究成果的能力，使它们能够实现范式转变，并在水管理研究计划和政策中考虑人类的影响。为了理解水与社会之间相互作用的动态性，应继续研究低成本、创新、可持续和社会可接受的技术，并协助开展收集和分发数据的工作。

1.6 阐述和 / 或强化科学知识、方法和工具，抵御涉水灾害，如洪水和干旱，实现及时预测。

approaches which assess hydro-climatic and non-climatic uncertainties integrated in hydrological projections and water use scenarios to support decision support system in water resources planning and management.

1.4. Conducting scientific research on the exploration of new business models, the role of water utilities, broadening engagement and partnerships, and infrastructure by the scientific community supported to accelerate the circular economy transition of the water sector.

83 Linear water management (Take-Use-Discharge) is still commonly adopted in the water sector and in the majority of watersheds today, and it is at the base of an unsustainable production and consumption pathways of water worldwide. The circular economy is an important part of integrated water resources management, also considering urban-rural interlinkages and the urban water systems within their biological regions.

84 The transition to a circular economy should consider the consumption and production of resources across the entire value chain, creating synergies within the water cycle for more just and efficient integrated water management. The use of a circular metabolic approach can further facilitate a more systemic and integrated water management across scales, particularly through the use of nature-based solutions, and appropriate consideration of integrated eco-system services, to better align the human water cycle with the natural water cycle, in enhancing and protecting resources, particularly facing climate mitigation and adaptation challenges.

85 IHP-IX will support such transition by researching several enabling factors including: system oriented scientific and technological advances, new business models and the role of water utilities, broad-based engagement and partnerships to favour a just transition, and re-thinking by design the retrofitting of existing and new infrastructures, to optimize water use and generate co-benefits, for example in relation to water-energy-food-ecosystem nexus. Existing infrastructure will need to be designed to fully enable resource efficiency and recovery and optimised to reduce energy consumption and decrease wastage, sharing infrastructure across sectors.

1.5. Undertaking and sharing assessments on the interaction between humans and water, in line with socio-hydrology by the scientific community supported to develop adaptive pathways, scenarios and strategies for water management.

86 The Water-human nexus has become very pertinent considering the impacts of changes due to anthropogenic activities on water resources. The continuum of human interactions with that of complex water management problems, lead us to many new questions and possibilities that hydrological scientists alone cannot address.

87 Social sciences play an increasingly important role for effective deployment of technology and methods involving “co-innovation” and “co-design” and are proving to be an effective manner to introduce new technologies in less developed regions and smaller villages.

88 Socio-hydrology provides two-way feedbacks between human and water systems that result in a wide range of phenomena that arise in different places around the world and in different contexts.

89 Supporting greater overlap in social and natural sciences creates better conditions for the design and implementation of projects to address complex issues such as the variability and change in the hydrological cycle under global changes and its social impacts. It provides a better foundation for decision-making in adapting to more devastating hydrological disasters, better management of the water-energy-food-ecosystem nexus, urban-regional water metabolism and better management of water scarcity and water systems, including transboundary water systems, as appropriate.

90 IHP-IX will facilitate the resolution of water-related societal problems by enhancing the understanding of the dynamics of water-societal interactions, underpinned by scientific findings supporting equitable solutions to achieve water security. IHP-IX will also support identifying synergies and trade-offs between societal goals related to water management. IHP will build capacity within Member States on the results of such research to enable them to move towards a paradigm shift and consider human influence in water management research plans and policies. Research should be pursued on the low-cost, innovative, sustainable, and socially acceptable technologies to address the understanding of the dynamics of water-societal interactions, assisting with data gathering and dissemination efforts.

1.6. Scientific knowledge, methodologies and tools in addressing water-related disasters, such as flood and drought elaborated and/or enhanced towards timely forecasting.

91 2018年水问题高级别小组最终成果文件《每一滴水都很重要》[4]表明：自1990年以来发生的1 000起最严重灾害中，涉水灾害占90%。在2019年联合国可持续发展高级别政治论坛上，联合国秘书长的报告强调，在最贫穷国家，由于灾害产生的较高经济损失率是消除贫穷的障碍。在“弥补差距和加速实施”这一章中，报告还强调“必须采用以人为本、确保全员参与的风险管理措施”。

92 需要采用多学科方法，以便更好地理解水文过程的这些变化。政府间水文计划第九阶段将加强和开发多学科知识库，以便更好地理解水文过程和极端现象的机制，分析水文气候变量的趋势，结合现场和遥感观测，解释气候模式预测结果。为了提高对水旱灾害的防御能力，政府间水文计划第九阶段战略规划将保持与世界气象组织的密切合作，进一步开发干旱和洪水早期预警系统和脆弱性评估的科学方法；为保障风险知情决策提供适当的工具和能力建设服务，加强成员国在地方、国家和跨界各级水旱综合管理方面的政策意识和体制能力建设。

1.7 支持科学界开发和共享全球变化和人类活动对河湖流域、含水层系统、沿海地区、冰冻圈和人类居住区影响的知识库，将其纳入水资源和服务管理计划。

93 需要开发用于评估过去、现在和将来从源头到海洋相互连接和轨迹变化的知识系统，并考虑冰冻圈、陆地水文水循环、地下水、泥沙和侵蚀过程以及沿海地区、三角洲和人类沿海住区的沉积物。河流、湖泊和含水层是景观的生命线，在水–能源–粮食–生态系统关系中具有核心功能，可以为人类提供饮用水、可再生能源或运输渠道，但同时也带来洪水风险和干旱隐患。它们形成了生物多样性的热点地区，即时反映了气候和土地利用的变化。同时，河流也因过度使用、泥沙连续性中断或空间限制而受到威胁。全球变化相关影响对河流和含水层等宝贵资源施加压力，不仅影响它们的数量，而且影响它们的质量。积雪和冰川的变化改变了积雪覆盖和冰川补给流域的径流和水资源的数量和季节性，进而对与全球生物多样性和生态系统服务相关的高山和低地生态系统产生了广泛影响。自然条件或采砂、采石等人类活动造成的侵蚀、泥沙输移和沉积过程对社会，尤其是在水资源可持续开发和管理方面，产生了重大影响。海平面上升可能导致海水入侵沿海含水层，影响地下水水质、污染饮用水水源并导致肥沃土地荒漠化。过度抽取地下水导致地面沉降，对人类居住区的基础设施产生了不利影响。

94 政府间水文计划第九阶段战略规划将开展对河流、湖泊和含水层从源头到海洋的相互连接和轨迹的综合研究和管理，重点关注河流水文学、水力学、泥沙动力学（包括侵蚀、输移和沉积作用）、地形动力学、地表水和地下水相互作用、包括在维护河流生态环境健康时的水质以及河流管理的基本过程。考虑社会经济挑战和全球变化驱动因素有助于改善水资源综合管理的一项成果将是成员国自愿参与完成的全球大河现状和未来情势综述。通过评估受气候变化影响的积雪、冰川和永久冻土的状况，加强科学家和机构之间的合作，并促进制定适应战略。通过继续促进在地方、区域和全球不同尺度上更好地了解泥沙启动、输移、储存和收支情况，支持有效、综合的泥沙管理。政府间水文计划第九阶段战略规划将继续开发地下水科学知识库，综合考虑全球变化的影响，促进对地下水资源及其所依赖的生态系统进行合理的、公平的管理。

4 联合国经济和社会事务部、世界银行，2018年。《每一滴水都很重要》：水行动议程–高级别水成果小组，华盛顿特区，第34页。

91 Water-related disasters accounted for 90% of the 1,000 most severe disasters that have occurred since 1990, according to the final outcome document 'Making Every Drop Count of the High-Level Panel on Water (2018)[4]. The Report by the Secretary-General of the United Nations at the 2019 UN High-Level Political Forum on Sustainable Development (HLPF)4 emphasizes that the higher ratio of economic loss caused by disasters in the poorest countries is an obstacle to eradicating poverty. The Report also highlighted that "all risk-management measures must be human-centered and ensure a whole of society approach" in the chapter of responding to gaps and accelerating implementation.

92 Multidisciplinary approaches are needed to better understand these changes in hydrological processes. The IXth phase of IHP will enhance and develop multidisciplinary knowledge base to better understand the mechanisms of hydrological processes and extremes and analyze trends of hydro-climatic variables and provide interpretation of climate model projections considering in situ and remotely sensed observations. In close collaboration with the World Meteorological Organization, IHP-IX will further develop scientific methodology on drought and flood early warning (EWS) systems and vulnerability assessment to increase resilience to floods and droughts; providing appropriate tools and capacity building to ensure risk-informed decision-making and allow Member States to strengthen policy and institutional capacity for integrated flood and drought management at the local, national and trans-boundary levels.

1.7. Development and sharing of knowledge-base on the impacts of global change and human usage on river and lake basins, aquifer systems, coastal areas, and cryosphere and human settlements by the scientific community supported so as to embed it in water resources and services management plans.

93 Knowledge systems need to be developed for assessing past, current, and future changes in the Source-to-Sea interconnection trajectory incorporating the cryosphere, the terrestrial hydrological water cycle, groundwater, sediment and erosion processes and deposition in littoral zones, deltas and coasts where numerous human settlements lay. Rivers, lakes and aquifers serve as lifelines in the landscape and have a central function in the water-energy-food-ecosystem nexus, supplying people with drinking water, renewable energy or transport means but are as well of central importance related to flood risk and droughts. They form the hotspots of biodiversity and reflect immediately climate and land use changes. At the same time rivers are endangered by overuse, interruption of sediment continuity or spatial restriction. Global changes related effects apply pressure on these valuable resources (rivers and aquifers) affecting not only their quantity, but also their quality. Changes in snow and glaciers have changed the amount and seasonality of runoff and water resources in snow dominated and glacier-fed river basins and have widespread consequences for high mountain and lowland ecosystems of global relevance for biodiversity and ecosystem services. The processes of erosion and sedimentation including sediment transport and sediment deposition both naturally or due to human activity, including sand and gravel mining, have many important implications for society, particularly in terms of the sustainable development and management of water resources. Sea level rise may lead to salt water intrusion into coastal aquifers affecting groundwater quality and contaminating drinking water sources and leading to desertification of fertile land. Land subsidence caused by excessive abstraction of groundwater effects negatively infrastructure in human settlements.

94 IHP-IX will provide an integrated river, lake and aquifer research and management in the source-to-sea interconnections and trajectory focusing on fundamental processes in river hydrology, hydraulics, sediment dynamics including erosion, transport and deposition, morphodynamics, surface water and groundwater interaction, water quality and river management incorporating socioeconomics challenges and global change drivers while maintaining the health of river ecological environment. As contribution to an improved integrated water management, one output will be a global overview of the status and future of large rivers with the participation of Member States on a voluntary basis. It will assess the state of the snow, glacier and permafrost impacted by climate change and strengthen cooperation among scientists and institutions and formulate adaptation strategies. It will continue to promote the development of an improved understanding of sediment mobilization, transport and storage and sediment budgets at local, regional and global scales, to support

4 UN Department of Economic and social Affairs and the World Bank. 2018. Making Every Drop Count: An Agenda for Water Action – High Level Panel on Water Outcome. Washington, D.C. 34 pp.

95 它将支持开展研究，以便更好地理解与应对全球城市和农村变化影响相关的水需求，以及综合城市水管理、水敏感城市设计及其他缓解方法和工具的作用。

1.8 支持科学界开发和分享关于改善水质和减少水污染的知识和创新解决方案，加强沟通，支持基于科学的决策，完善知识和服务，降低健康风险。

96 必须加强世界水资源状况和健康的知识库建设。需要在流域、国家和区域层面开展综合水质评估，以支撑水和卫生设施管理和政策优先事项，从而提升和恢复水质。需要开展研究，以便加强气候变化对水质影响的科学认知和知识，这是一个缺乏数据和科学信息的研究不足的领域。需要开发和推广用于监测和评估水质的创新工具，并将其与卫生倡议相关联。为了制定有效水质监测、评估和管理战略，需要卫生、水质条例、标准和规范等科学基础。

97 政府间水文计划第九阶段战略规划将继续支持综合水质评估，强化基于生态系统的水质管理方法的知识和研究，特别是更好地了解与水质有关的生态系统产品和服务的变化，例如水质恶化造成的生态系统退化和生物多样性丧失。指导决策者利用这些知识，评估与水质有关的生态系统产品和服务，并制定水质恢复战略。

1.9 交流和 / 或使用由科学界和服务供应商开发和分享的地球观测、人工智能和物联网等新技术，加强涉水利益相关方在水文规划、评估以及监测和站网布设方面的能力建设。

98 信息通信技术创新和人工智能相关技术似乎层出不穷，促进有效且高效地利用水资源，有助于实现若干可持续发展目标，如获得饮用水、环境卫生和个人卫生、减灾。

99 通过将传感器接入个人移动设备和物联网，开发和实施用于观测、获取和分发数据的新一代全球网络。越来越多的地球观测卫星也提供了大量的水文信息。许多相关事项，如及时灾害预测、地下水治理、纳米卫星使用、循证规划、实时监测、优化使用资源和时间的有效决策支持系统等，都将受益于新技术。

100 政府间水文计划第九阶段战略规划将努力支持并实地测试这些新技术，验证它们在为子孙后代改善可持续水管理和保护生态系统方面的工作原理。特别在发展中国家，人员和机构能力仍是推进电子学习和数字化的主要障碍。政府间水文计划第九阶段战略规划将尽最大努力建立一个实践社区，实现电子化教育内容的共享。

1.10 支持科学界和其他利益相关方开展并分享在水文学科中融入公民科学的研究，提高对水循环的认识，促进基于科学的决策。

101 挖掘公民科学潜力，对水文科学和水资源管理中传统的科学数据收集和知识生成方式进行补充，在创造新的水文知识，特别是有关水文过程特征描述、异质性、偏远地区和人类对水循环影响这四个方面有巨大前景。如果按照科学有效的方法开发了有效工具，对数据进行抓取、整理、发布和质量控制，那么公民科学家就可通过卫星、低成本传感器、嵌入网络地图工具的智能手机，以及与社交网络服务器交换信息的全球卫星导航系统，利用生成的涉水信息开展水资源管理。因此迫切需要开发和应用人工智能等恰当的技术，融合物联网、遥感和公民科学项目获得的不同来源的数据。

effective and integrated sediment management; IHP-IX will continue developing a scientific knowledge base on groundwater integrating considerations of global change effects for a rational and equitable management of the groundwater resources that include dependent ecosystems.

95 It will support research to better understand the water demand associated with the response to the effects of global changes in both urban and rural context and the role of Integrated Urban Water Management (IUWM), Water Sensitive Urban Design (WSUD) and other approaches and tools to mitigate them.

1.8. Development and sharing of knowledge and innovative solutions on improving water quality, and reducing water pollution by the scientific community supported and communicated to support science-based decision-making, improve knowledge, services and reduce health related risks.

96 It is essential to strengthen the knowledge base on the state and health of the world's water resources. There is a need for comprehensive water quality assessments at basin, national and regional levels in order to underpin water and sanitation management and policy priorities to improve and restore water quality. Research is needed to enhance the scientific understanding and knowledge on the impact of climate change on water quality, which is an under-researched area where data and scientific information are lacking. Innovative tools to water quality monitoring and assessment need to be developed and promoted and linked with sanitation initiatives. The scientific underpinning of sanitation, water quality regulations, standards and criteria is necessary to develop effective water quality monitoring, assessment and management strategies.

97 IHP-IX will continue supporting comprehensive water quality assessments and will enhance knowledge and research promoting an ecosystem-based approach to water quality management, in particular to better understand changes in water quality-related ecosystem goods and services such as the ecosystem degradation and biodiversity loss caused by water quality deterioration. This knowledge will be used to contribute in providing guidance to policy-makers in valuing water quality-related ecosystem goods and services and in developing water quality restoration strategies.

1.9 Development and sharing of new technologies using earth observation, Artificial Intelligence and Internet of Things by the scientific community and service providers are communicated to and/or used for capacity strengthening of water stakeholders to increase their use in hydrological planning and assessment as well as monitoring and distribution networks.

98 There are seemingly an unending number of Information and Communication Technologies (ICT) innovations and Artificial Intelligence (AI)-related technologies impacting efficient and effective use of water resources and realizing several SDGs related to access to drinking water, sanitation and hygiene, and disaster mitigation.

99 The inclusion of sensors to personal mobile devices and the Internet of things (IoT) enables us to develop and implement a new generation of observation, data-acquisition, and data-distribution networks globally. The increasing number of Earth Observations satellites also provide large pool of information on hydrology. Many related issues such as timely disaster forecasting, groundwater governance, the use of CubeSats (nano-satellites), evidence-based planning, real-time monitoring, and effective decision support systems optimizing the use of resources and time, will all benefit from new technologies.

100 IHP-IX will work to support and field-test these advances and how they can improve sustainable water management for future generations and the preservation of ecosystems. Human and institutional capacity remains the major bottleneck to advancing e-learning and digitally supported particularly in developing countries. IHP-IX will maximise its efforts to build a community of practice and enable the sharing of digital educational e-content.

1.10 Conducting and sharing of research on integrating citizen science in the hydrological discipline by the scientific community and other stakeholders supported, to improve understanding of the water cycle enabling science-based decision making.

101 Exploring the potential of citizen science to complement more traditional ways of scientific data collection and knowledge generation for hydrological sciences and water resources management might have a significant potential to create new hydrological knowledge, especially in relation to the characterization of hydrological processes, heterogeneity, remote regions, and human impacts on the water cycle. Water-related information generated though satellites, low cost sensors and smartphones embedded with web-based mapping tools and global satellite navigation systems exchanged on Social Networking Services (SNSs) by citizen scientists

102 公民科学已成为开展水文研究的工具之一，汇聚科学家和公民收集的研究数据，为科学家决策提供支撑。包括虚拟领域在内的用户友好型技术的进步促进了交流、培训、在线数据的可视化以及数据收集工作。从科学的角度来看，公民科学扩大了数据收集可能性的时空范围，特别是在地方尺度上。

103 已有许多公民科学倡议和研究项目，可通过开放型科学和开放存取倡议为已有大数据添砖加瓦。但是，对公民科学数据准确性和质量的担忧影响了这些数据的全面认可度。有必要进一步阐述造成公民科学数据质量问题的机制和源头，并为提高此类数据的准确性和质量提供验证机制和指南。

104 政府间水文计划第九阶段战略规划将支持开发科学有效的方法和工具，助力包容性知识的生成过程，如公民对科学研究的贡献。

105 因此，政府间水文计划第九阶段战略规划将创造有利环境，通过加强水知识和水教育计划，协助公民和科学家使用科学方法，参与研究并汇报成果，以便增加公民科学对水文学研究的贡献。培训将特别有助于提高数据的准确性和有效性。此外，应该开发一些科学工具，以便鼓励公民参与水管理并且能够改善水管理的其他社会应用，如将现代科学与祖先、本土和地方知识相结合。

与其他联合国机构和科学伙伴合作，引领科学、研究和创新

106 为了履行其科学使命，政府间水文计划第九阶段战略规划将在前几个阶段成果基础上，引领促进可操作的科学研究和创新，以便应对复杂的相互关联的水挑战。它将继续促进科学合作，并与国际水文科学协会、国际水文地质学家协会、国际水利与环境工程学会等专业科学组织和其他国际科学水计划以及其他联合国组织建立伙伴关系，例如，与世界气象组织在监测和水文预报方面开展合作，与联合国环境规划署在互补和协同的水质方面开展合作，与联合国人居署和世界卫生组织在城市水挑战方面开展合作，与国际湿地组织和拉姆萨公约在湿地方面开展合作。此外，它将继续与学术机构和研究中心合作，制定研究计划，验证研究成果，并向成员国传播研究成果。它将继续扩大和深化不同的合作战略和协调战略，把水文学和其他学科连接起来，以便应对全球水挑战。

二、优先领域 2：第四次工业革命中可持续发展水教育

107 《2030 年可持续发展议程》和涉水可持续发展目标及其子目标的成功与否取决于人类价值观的深刻转变，以及直接影响人类生活方式的人类行为和行动的改变。只有当社会认识到需要将自身与自然重新结合，并对自然资源基础对提高人类生活质量的重要性和局限性这一普遍认知达成共识时，才能实现设想目标。

108 在全球变化背景下，为了提升水文化，在各成员国开展不同层级的水教育无疑是一个强大手段，通过践行包容性的循证水治理和管理，朝着韧性和可持续社会发展。水教育鼓励社会各界采取与生态系统再生速度相一致的可持续消费和生产模式。

109 因此，人类必须从生命早期阶段开始接受水教育，随后在所有年龄段和所有社区继续以各种方式接受水教育，树立水管理意识，唤醒公民对其权利义务的批判和解放意识，使他们成为积极的公民。我们必须拥有一批新型科学家、规划师和从业人员，他们具备解决复杂的、相互关联的水挑战的适当技能，准备在这十年结束前承担第四次工业革命中水务部门的职责。

can contribute to water resources management, if effective tools are developed following scientifically valid methodologies to capture, organize, quality control, and make such data available. There is an urgent need for the development and application of proper techniques, including AI, that can merge different sources of data obtained from IoT, remote sensing, and citizen science projects.

102 Citizen science has become one of the tools for hydrological research, enabling the efforts of scientists and citizens to collect data for research to be interpreted by scientists for decision-making. Advances in user-friendly technology including those in the virtual arena, also facilitates communication, training and online data visualization and data collection. From a science perspective, citizen science widens spatial and temporal data collection possibilities, particularly at the local scale.

103 Many citizen science initiatives and research projects already exist and may add to the big data available already through open science and open access initiatives. Concerns about the accuracy and quality of data generated through citizen science hamper full acceptance of such data. There is a need to further elaborate on how and where quality problems in citizen science data can arise, on validation mechanisms and on guidelines, in order to improve the accuracy and quality of such data.

104 IHP-IX will support the development of scientifically valid methods and tools that promote inclusive knowledge generation processes, such as citizens' contribution to scientific research.

105 IHP-IX will therefore create the enabling environment and assist citizens and scientists, through enhanced water knowledge and education programmes to ensure scientific methods are used when participating in and reporting their findings to increase the contribution of citizen science to hydrology research. Training, in particular, will contribute to enhancing accuracy and validity of data. Additionally, scientific tools should be developed to encourage citizen participation and other social applications that can improve water management, such as integrating modern science with ancestral, indigenous and local knowledge.

<u>Leading science, research and innovation in cooperation with other UN Agencies and scientific partners</u>

106 In line with its Mandate on science, IHP-IX will provide leadership in advancing actionable scientific research and innovation for addressing complex interlinked water challenges building on achievements from previous phases. IHP-IX will continue to promote scientific cooperation and building partnerships with other UN organizations, such as WMO on monitoring and hydrological forecast and UNEP on water quality for complementary and synergy, UN-HABITAT and WHO on urban water challenges, Wetlands international and the Ramsar convention on wetlands, and with professional scientific organizations such as IAHS, IAH, IAHR and other international scientific water programmes. Additionally, IHP- IX will continue partner with academic institutions and research centres in developing research initiatives, validating results and disseminating these results to Member States. IHP-IX will continue to broaden and deepen different collaboration and coordination strategies, working on interfaces between hydrology and other disciplines to address the challenges.

PRIORITY AREA 2 : WATER EDUCATION IN THE FOURTH INDUSTRIAL REVOLUTION INCLUDING SUSTAINABILITY

107 It is undeniable that the success of Agenda 2030 for Sustainable Development and water-related SDGs and associated targets depends on a profound transformation in human values and, consequently, human behaviour and actions, directly impacting how we live our lives. Achieving that end can only be envisioned when society recognizes the need to reintegrate itself with nature in ways that embrace a common understanding of the importance and limits of our natural resource base to improving our quality of life.

108 Water education at all levels for an improved water culture, in a context of global change, is undoubtedly a formidable tool for Member States to practice inclusive, evidence-based water governance and management in order to move towards resilient and sustainable societies. It is a tool that encourages the engagement of all sectors of society to adopt sustainable consumption and production patterns that are in tune with the regeneration pace of ecosystems.

109 Water education must therefore begin at an early stage in life and continue to be offered in a variety of ways to build a water stewardship mentality at all ages and in all communities, awakening critical and emancipatory awareness in citizens in relation to their rights and duties so that they can be active citizens. We must have a cadre of new scientists, planners, and practitioners equipped with appropriate skills for addressing complex interconnected water challenges and ready to assume positions of responsibility in a fourth industrial

110 到 2029 年，全球将对大量决策者、教育者和公民进行培训，并依靠健全的科研信息提高他们对涉水挑战和机遇的认识，丰富他们的知识，从而促进可持续水管理和治理。并加强科学家网络，以编制和传播相关材料，举办培训 / 提高认识活动。

1 优先领域 2 与《2030 年可持续发展议程》之间的关系

111 水教育与可持续发展目标 6 及其子目标（确保所有人对水和卫生设施的使用和其可持续管理），以及所有其他涉水可持续发展目标有关，因为实现这些目标需要训练有素和有可持续发展意识的人力资源。

112 可持续发展教育使每个人都能获得塑造可持续未来所需的知识、技能、态度和价值观。这也与可持续发展目标 4（确保包容性和公平性优质教育，扩大全民终身学习机会），特别是其子目标 4.7（确保所有学习者获得促进可持续发展所需的知识和技能）有直接联系。

113 同样，鉴于广为流传的“蓝线”概念以及可持续发展目标中的所有涉水目标，优先领域 2 与可持续发展目标 9（工业、创新和基础设施）的各子目标相关。它与可持续发展目标 12（可持续消费和生产模式）及其子目标的行为和制造业转型方面也直接相关。

114 此外，政府间水文计划成员国认识到《2030 年可持续发展议程》中的一个空白领域，已要求秘书处与世界卫生组织、联合国环境规划署和经济合作与发展组织合作，根据可持续发展目标的子目标 6a，开发和使用一个与水教育有关的指标。

2 预期成果

2.1 提高各级公众对水在家庭生活、生态系统和生产发展中具备多重功能的认识，更好地了解他们在其中的重要贡献。

115 变革是困难的，可能受到诸如人类抵制或传统社会风俗、资金或技术等方面的多种阻碍。很明显，如果没有社区的理解和支持，任何提议解决方案都不会完全有效。对于传统社会和那些受到冲突或任何环境或个人压力胁迫的社会来说尤为如此；他们的日常生活模式在逻辑上集中于日常生存策略，进一步加剧了变化的复杂性。

116 许多人对于他们日常生活与供水之间的关系只有模糊的认识，经常认为水是理所当然的东西。有效教育能引导公众更好地理解并随后应用新知识；它直接或间接地影响那些由于采取负责任立场而提高了认识的人，所以，教育的结果是变革。必须强调妇女、年轻人和青年专业人员是改善水管理和治理的变革推动者，通过参与制定创新的科学计划，能使他们了解到水在生活中的重要性，以确保做好未来几代水领袖的人才储备工作。因此，在涉水决策过程中应积极征求年轻人的意见。

117 因此，必须强调水教育应包含战略和技术内容，这些战略和技术将帮助人类提高认识，从而采取更好的做法来保护水资源和生态系统，并保障水资源可持续利用。那些认识到水资源对人类生计、经济发展和环境保护具有价值的人们，将会要求保护水资源以促进其可持续管理和治理。

118 可通过终身学习、以祖传知识形式讲述社区故事、实地讲习班培训、交流项目、进修课程、暑期学校、研究生学位和社会媒体等多种途径获取与水有关的知识。此外，新冠疫情引起了基于技术型工具的教

revolution setting in the water sector by the end of this decade.

110 By 2029, a critical mass of decision makers, educators and citizens worldwide will be trained, have their awareness raised and their knowledge enriched on water-related challenges and opportunities based on sound scientific and research information to facilitate sustainable water management and governance and governance. Networks of scientists will be strengthened to develop and disseminate related material and conduct the training / raising aware sessions.

Relation between Priority and the Agenda 2030

111 Water education is connected to SDG 6 and all of its targets (ensuring the availability and sustainable management of water and sanitation for all) and all other water-related SDGs' targets, as they all require trained and aware human resources to be achieved.

112 Education for Sustainable Development allows every human being to acquire the knowledge, skills, attitudes and values necessary to shape a sustainable future. There is also a direct link to SDG 4 on ensuring inclusive and equitable quality education and promote lifelong learning opportunities for all and especially to the target 4.7 to ensure that all learners acquire the knowledge and skills needed to promote sustainable development.

113 Similarly, this Priority Area is related to various targets of SDG 9 (industry, innovation and infrastructure), and given the widely circulated "blue thread" concept, as well as water-related targets of all of the SDGs. The behavioural and manufacturing transformation aspects of SDG 12 and its targets (sustainable consumption and production patterns) are also directly connected with this priority area.

114 Additionally, the Member States of the IHP recognizing a gap in the 2030 Agenda, have requested the Secretariat to pursue in cooperation with WHO, UNEP and OECD the development and use of a water education-related indicator under target 6a.

Expected outputs:

2.1 Public's awareness at all levels raised towards better understanding their contribution to the important multi-functions of water in domestic life, ecosystems and productive development.

115 Change can be difficult and impeded by any number of obstacles, be it human resistance or customary social mores, financial or technological impediments, among others. It is clear that no proposed solution will be fully effective if imposed on a community without their understanding and support. This is particularly the case for traditional societies and those under duress from conflicts or any number of environmental or personal stressors; their daily life patterns being logically focused on daily survival strategies and further complicating change.

116 Many people have only a vague understanding of the relationship of their daily lives with the availability of water and often consider water for granted. Effective education leads to better understanding and subsequent application of new knowledge by the public, either directly or indirectly to those influenced by those whose awareness was raised, as they take positions of responsibility; change being the output of such a process. It is important to emphasize the contributions of women, youth and young professionals as agents of change to improving water management and governance by gaining an understanding of the importance of water in their lives through becoming involved in the development of innovative science programmes to ensure that future generations of water leaders are in the making. Therefore, the opinions of young people should actively be sought-out as inputs to decision-making processes related to water.

117 It is therefore crucial to emphasize that water education must embrace strategies and techniques that will support people to enhance their awareness and be able to adopt better practices towards protection and sustainable water use and ecosystems. People who are aware of the value of water for their livelihoods, economic development and protection of the environment, will demand its conservation contributing to its sustainable management and governance.

118 There are many ways to acquire water-related knowledge including among others lifelong learning, community story-telling as a form of ancestral knowledge, training in-field workshops, exchange programmes, refresher courses, summer schools, graduate degrees and social media. Additionally, the COVID-19 pandemic caused a profound change in educational modes based on technological tools. IHP will continue to utilize the new possible modalities borne out of this necessity to increase its reach and benefit all people, urban and rural, in attaining a more profound understanding of their dependency on and relationship to water resources while

育模式的深刻变化。考虑到人们接入互联网机会的不均等性，政府间水文计划将继续利用由疫情带来的知识传播新模式来扩大水知识的覆盖面，使所有城市和农村人口受益，让他们更深刻地了解自己对水资源的依赖及其与水资源的关系。

119 政府间水文计划第九阶段战略规划将继续推广水教育的广义概念，同时通过提高认识的活动在成员国监管框架内创造条件，促进行为转变以加强生态社会意识。它将支持诸如指南、简报和水教育最佳实践案例研究等跨学科材料编制，供大众媒体用于提高广大公众的认识。它将利用联合国教科文组织指定地点和水博物馆网络，提高公民意识、提升公民水文化。政府间水文计划第九阶段战略规划将与水青年网络合作，加强作为变革推动者的青年的能力，并促进他们参与水资源决策过程。

2.2 促进联合国教科文组织水事机构发展和实施跨学科研究合作和教育，加强全面参与性实践。

120 将研究与教育举措相结合，既培养研究能力，又促进公众对水科学的支持和理解。从本质上讲，公共领域的水科学宣传是一项涉及各种支持者和利益相关方的合作事业。而且，在一个日益全球化和多样化的世界里，让水科学部门在开发技能、培训领导者、提高公众意识以及转让技术和技能知识方面发挥推动作用，是实现可持续水安全未来的关键。展览、通俗易懂且易于获取的出版物、在线平台以及其他公共事件等活动都是扩大和增强水科学影响的重要途径。

121 在联合国教科文组织秘书处和政府间水文计划的支持下，联合国教科文组织水事机构完全有能力参与公共宣传和教育，使水科学的社会影响多样化。联合国教科文组织水事机构由水科学家和社会科学家组成，便于与联合国教科文组织其他部门、政府间水文计划成员和政府实体合作，促进公众参与生态水文实践和学习。

2.3 为各级正式、非正式和非正规教育编写与水有关的教学材料，使人们更好地了解水在生活和社区中的重要性。

122 教育通常是在正式、非正式和非正规环境下进行的。无论采用何种课程和授课机制，水教育必须以高质量的科学为基础，才能对改善水管理和治理产生最大的影响。对整个教育过程而言，高质量的科学、可靠的数据和技术以及教育者 / 培训者交流这些信息的能力之间相互关联至关重要。此外，教育需要采用最相关的技术，以确保教育成果的质量，并使教育成果惠及所有人；随着时间的推移，形成一种积极主动的代际水管理环境。因此，具有强大科学基础的广泛水教育战略是塑造每个人未来水意识的决定性因素。

123 促进思维进化最有效的方式是对社会各部门开展教育，使人们更多地了解水在每个人生活中的作用。伴随这一必要转变，社会正经历第四次工业革命，其特征是在生物技术、大数据、无人机和人工智能等领域涌现了大量新技术，这些技术将重塑经济、研究和专业水实践。因此，水教育必须利用这些技术帮助专业人员和技术人员做好准备，作出最佳管理决策，并更加聚焦必要的研究和能力建设活动。

124 一个关键挑战是利用新技术和创新学习过程来开发最先进的培训计划和材料，如以简短教学视频、电子教室和会议等形式进行的（开放式）电子学习课程，甚至可能包括在线研究生学位课程。

125 联合国教科文组织在水教育领域有着悠久的历史，它支持开放式教育资源计划，还支持通过职业和高等教育与研究来获取新的水知识。

keeping in mind the inequality of internet access.

119 IHP-IX will continue to encourage a broad conception of water education, along with conditions in the regulatory frameworks of the Member States that favour a change in behaviour towards a society with greater eco-social awareness via awareness raising activities. IHP-IX will support the development of interdisciplinary materials, such as guidelines, briefing papers, and case studies on leading practices in water education for mass media contributing to raising awareness of public at large. UNESCO-designated sites and the network of water museums will be mobilized in raising awareness and improving water culture of citizens. IHP-IX in partnership with water youth networks will strengthen the capacities of youth as agents of change and promote their involvement in water decision-making processes.

2.2 Development and implementation of transdisciplinary research collaborations and educational approaches by UNESCO Water Family promoted to enhance participatory holistic practices.

120 Connecting research to educational initiatives builds both research capacity and promotes greater public support and understanding of water science. Water science advocacy in the public realm is inherently a collaborative undertaking involving a variety of constituencies and stakeholders. Moreover, in an increasingly globalized and diversifying world, putting the water science sector to work in advancing the development of skills, training leaders, growing public awareness, transferring technologies and technical knowledge, is critical to achieving a sustainable and water secure future. Activities, such as exhibitions, jargon-free and easily accessible publications, online platforms, and other public events all offer important avenues to broaden and enhance the impact of water science.

121 With the support of the UNESCO Secretariat and IHP, the UNESCO Water Family is well positioned to engage in public outreach and education that diversifies the social impact of water science. The UNESCO Water Family consists of both water and social scientists and as such it is well positioned to collaborate with other UNESCO divisions, members of the IHP, and governmental entities to increase participatory ecohydrological practices and learning.

2.3 Teaching and learning materials on water-related matters for formal, non-formal and informal education at all levels elaborated towards a better understanding of the importance of water in lives and communities.

122 Education is normally delivered in formal, non-formal and informal settings. Regardless of the delivery context——curricula and transfer mechanism employed——in order for water education to have the greatest impact on improving water management and governance, it must be based on quality science. The linkage between quality science, credible data and technology and the ability of educators/trainers to communicate such information is fundamental for all education processes. Additionally, education needs to employ the most relevant technology and to ensure the quality of outputs as well as reaching all people; over time engendering a pro-active and inter-generational water stewardship context. Therefore, a broad water education strategy with a strong scientific basis is a determinant factor to shaping a water conscious future for everybody.

123 The most efficient way to catalyse this evolution in thinking is through education to all sectors of society leading to a greater understanding of the role that water plays in every individual's life. Along with this needed transformation, our society is experiencing a fourth industrial revolution, characterized by the emergence of a new broad range of technologies in fields like biotechnology, big data, drones and artificial intelligence, among others, that will reshape the economy, research and professional water practice. Hence, water education must use those technologies to help prepare professionals and technicians to make the best management decisions and to better focus needed research and capacity-development activities.

124 A key challenge is developing state-of-the-art training programmes and materials using new technologies and innovative learning processes, such as (open) e-learning in the form of short instruction videos, e-classrooms, and meetings, even potentially including on-line graduate degree programmes.

125 UNESCO has a long history in the field of water education including support to Open Educational Resources (OER) programmes as well as in professional and tertiary education and research to garner new water knowledge.

126 政府间水文计划第九阶段战略规划将发动联合国教科文组织水事机构，并与联合国教科文组织其他部门特别是教育部门合作，将聚焦所有教育类型的水教育和学习材料的设计、规划和实施，包括改进K–12课程[5]中关于水问题的教学工具。

2.4 支持科学界基于新实践来开发和分享方法与工具，将科学信息转化为促进教育、决策和政策制定的格式。

127 科学信息的可及性和可见性是开放型科学的先决条件。数据一旦被处理成科学信息并发表在期刊上，即需要共享和传播，供公民、专业人员、科学家和官方使用。应将科学信息与本土/地方知识相结合，并通过科学期刊、教育资源和其他广泛咨询的媒体和数字渠道广泛传播。

128 一般来说，目前将科学信息转化为决策和政策制定格式的方法并不多，如可视化方法、有助于启发决策或情景规划的路线图等。因此，有必要让流域一级的所有利益相关方参与开发新思路，并通过多媒体传播新方法。

129 政府间水文计划第九阶段战略规划将协助开发和传播数据可视化新方法，以便促进基于科学的决策并提高公众意识。

2.5 加强与水有关的高等教育和职业教育中熟练专业人员和技术人员的能力，识别可持续水管理的主要差距，为政府和社会消灭这些差距和实现《2030年可持续发展议程》目标提供合适的工具。

130 鉴于社会面临的涉水问题的复杂性，提高水项目和培训人员的数量和质量应成为所有管辖级别的高度优先事项。

131 许多水技术人员和专家、教师、年轻专业人员和教授需要通过在职培训来提高他们在野外、实验室和课堂技能方面的能力，使他们能够在《2030年可持续发展议程》背景下更有效地执行任务。

132 相关培训水利技术人员的涉水职业课程已逐步递减，而技术的快速转型和创新则要求加强水务部门现有技术型人力资源。需要通过在联合国教科文组织水事机构内以及与联合国系统其他机构和计划开展合作，保持并扩大对涉水领域（如：水文气象监测、灌溉系统、卫生设施、供水系统）技术人员的培训。为了应对复杂和相互关联的水挑战，加快落实可持续发展目标6和其他涉水可持续发展目标，成员国需要加强高等水教育，目的是培训科学家以便进一步发展水科学，并教育新一代水利专业人员、管理人员和决策者。

133 政府间水文计划第九阶段战略规划将动员和加强联合国教科文组织水事机构、世界水资源评估计划、联合国教科文组织教席和二类中心（如：荷兰代尔夫特国际水利环境工程学院、国际理论物理中心）之间的伙伴关系，开展满足成员国需求的前沿教育计划。

134 政府间水文计划第九阶段战略规划将开发/支持与水教育、支持提高职业和高等水教育能力有关的跨学科材料编制，这些材料包括指导方针、简报、课程、原型专业发展计划、案例研究和最佳实践，特别是增加发展中国家的正规和非正规水教育工作者/培训师和拥护者的数量，同时培养他们对国家水环境和地方需求的敏感力。

5 字母K代表幼儿园，12代表年级。K–12涵盖了从幼儿园到12年级的时间，包括幼儿园、小学、中学、高中和大学预科教育。

126 IHP-IX will mobilize the UNESCO Water Family and will collaborate with other parts of UNESCO, particularly the Education Sector and will focus on the design, planning and implementation of teaching and learning materials on water for all types of education, including the development of improved tools for the teaching of water issues in the K-12[5] curriculum.

2.4 Development and sharing of methods and tools based on new practices by the scientific community supported to translate scientific information into a format facilitating education, decision-making and policy formulation.

127 Accessibility and visibility of scientific information are prerequisites for open science. Once data has been processed into scientific information and published in journals, it needs to be shared and disseminated, allowing it to be used by citizens, professionals, scientists, and authorities. Scientific information should be combined with indigenous/local knowledge and widely disseminated in scientific journals, education sources and other widely consulted media and digital outlets.

128 The current methods for translating scientific information into a format for decision-making and policy formulation, such as visualization methods, roadmaps that provide implications for decision-making or scenario development are, in general, limited. Therefore, it is necessary to develop new ideas, disseminate new methods through multiple media and involve all stakeholders at the basin level in this process.

129 IHP-IX will assist in development and dissemination of new data visualization methods to facilitate science-based decision making and public awareness.

2.5 Capacities of skilled professionals and technicians at water-related tertiary and vocational education strengthened to identify the main gaps for sustainable water management towards providing appropriate tools to governments and societies to address those gaps and the Agenda 2030 targets.

130 Given the complexity of water-related issues confronting society, increasing the number and quality of water programmes and trainers should be a high priority at all jurisdictional levels.

131 There are numerous water technicians and experts, teachers, young professionals and professors who require on-the-job training that will improve their capacity in the field, lab and classroom skills and enable them to perform their tasks in a more effective manner in the context of Agenda 2030.

132 The offers of water-related vocational programmes to train water technicians has declined steadily and the rapid transformation and innovation of technologies require the enhancement of existing human resources at the technical level in the water sector. Within the UNESCO Water Family and in partnership with other UN system agencies and programmes, efforts are needed for maintain and expand the training of technicians in water-related fields (e.g. hydrometeorological monitoring, irrigation systems, sanitation, water supply systems). To address complex and interconnected water challenges and accelerate implementation of SDG 6 and other water-related SDGs, Member States will need to enhance tertiary water education aiming at the training of scientists to further develop water sciences, as well as to educate the new generations of water professionals, managers, and decision makers.

133 IHP-IX will mobilize and strengthen partnerships among the UNESCO Water Family, WWAP, UNESCO Chairs and Category 2 Centres such as IHE-Delft and also ICTP (International Centre for Theoretical Physics) for cutting-edge educational programmes that meet the needs of the Member States.

134 IHP-IX will develop/support development of interdisciplinary materials, such as guidelines, briefing papers, curricula and prototype professional development programmes and case studies and best practices connected with water education and support enhancement of vocational and tertiary water education capacities, particularly in developing countries by increasing the numbers of formal and informal water educators/trainers and champions, being sensitive to national water contexts and local needs.

5 The letter K stands for Kindergarten and 12 for 12th grade. K–12 covers the years from Kindergarten through 12th grade and includes Kindergarten, Primary, Secondary, High–School and Pre–University education

2.6 加强决策者、水管理者和主要水务机构部门的能力，使他们能够利用新技术和研究，更好地决策、设计和实施综合、高效的水政策。

135 尽管自下而上的方法对提高人们的认识非常重要，但关键是要认识到政府或大公司等决策者的行动对社会和环境的影响最大。因此，需要实施几种不同的战略，以满足活跃在社区决策过程中的所有部门的需求和利益。好的政策需要知情的公民，而好的决策需要健全的科学，这反过来又需要拥有一系列自然、技术和社会科学学科知识的专家和机构。为了使决策者能够依靠必要的社会支持来设计和实施高效和可持续的水政策，必须根据他们自身需求以及信息和知识对水治理和水管理的影响，来定制专门的培训，这是至关重要的。

136 这套方法将为决策者和机构提供必要的工具，促进从单一的、基于消费的经济向基于管理、可持续和保护的经济转型。

137 政府间水文计划第九阶段战略规划将动员联合国教科文组织水事机构编制适应性的培训材料，并为跨学科支持材料编制提供技术援助，这些支持材料如指导方针、信息文件以及关于决策者和水管理者能力建设的先进做法的案例研究。在联合国可持续发展目标 6“全球加速框架”下，政府间水文计划第九阶段战略规划将与联合国水机制共同领导、支持各国能力建设。

3 与其他联合国机构和合作伙伴合作，协调水教育工作

138 作为联合国的主要教育组织，联合国教科文组织将通过它的水事机构与教育部门伙伴共同履行使命，领导水教育优先领域，支持开发从小学到研究生教育的水教育课程、终身学习课程以及正式、非正式和非正规的水教育活动。除了在联合国可持续发展目标 6“全球加速框架”下与联合国水机制共同领导、支持成员国的能力建设外，联合国教科文组织还将与其他联合国机构以及包括涉水非政府组织、专业协会及私营部门在内的其他相关伙伴合作，开展互补或联合培训活动（例如，与世界气象组织合作开展国家水文服务的能力建设）。

三、优先领域 3：弥合数据与知识之间的鸿沟

139 联合国教科文组织即将兑现的一个承诺是：使数据透明度和可获取性成为持续促进开放型科学的主要支柱之一。水文测量对于决策和可持续水资源管理至关重要。缺乏或无法访问关于水资源的数量、质量、分布、获取、风险和使用等方面的全面或长期数据，往往会导致部分管理和投资失效。因此，需要确保多数情况下改善数据的充足性和可获取性。

140 水数据有各种来源，且数据生成器具有多样性。然而，收集和理解原始数据并将其应用到决策环节的水文系统往往比最初预期的要复杂得多。只有以透明、可理解的方式来收集数据，并详细描述所需解决的具体问题，才能弥合数据与知识之间的鸿沟。需要在濒河国家之间建立适当的数据网络，向所有感兴趣的用户提供跨界水资源数据。收集、共享和解释跨界水资源数据所面临的挑战变得更加复杂，因此，有必要加强数据收集，以确保数据质量、填补地方监测数据空白以及提升监测信息价值。

141 弥合数据与知识之间的鸿沟要求提供、整合和处理不同学科和来源的数据；政府间水文计划第九阶段战略规划将充分利用联合国教科文组织水事机构现有数据和知识库，如联合国教科文组织二类中心——国际地下水资源评估中心开发的全球地下水信息系统（创建于 2004 年）。

2.6 Capacities of decision makers, and water managers and key water sector institutions strengthened allowing them to take advantage of new technologies and research to enhance better decisions, design and implementation of integrated and efficient water policies.

135 Even though bottom-up approaches are very important to raise awareness among people, it is crucial to acknowledge that the actions of the decision makers such as governments or big companies have the biggest impacts on society and environment. Therefore, several different strategies need to be implemented to address the needs and interests of all sectors active within community decision-making processes. While good policy requires informed citizens, good decision-making requires sound science, which in-turn requires knowledgeable experts and institutions in a range of natural, technological and social science disciplines. For decision-makers to be able to count on the social support necessary to design and implement efficient and sustainable water policies, it is essential that they have access to specialized training, based on their needs and the impact that information and knowledge can have on the governance and management of water.

136 This suite of approaches will provide decision-makers and institutions with the necessary tools for boosting the transition from an economy based solely on consumption to an economy based on stewardship, sustainability and conservation.

137 IHP-IX will mobilize the UNESCO Water Family in developing adapted training materials and providing technical assistance for the development of interdisciplinary support materials, such as guidelines, information documents, and case studies on leading practices in capacity building for decision makers and water managers. IHP-IX will co-lead the effort of UN-Water on providing capacity development support to countries under UN SDG 6 Global Acceleration Framework.

Coordinating Water Education in cooperation with other UN Agencies and partners

138 In line with its mandate as the principal UN educational organization, UNESCO through its Water Family and in partnership with the Education Sector will provide leadership on this priority area by providing support on curriculum development, life-long learning and formal, informal and non-formal water education activities from primary schools to post-graduate education. In addition of co-leading the UN-Water efforts on capacity development under UN SDG 6 Global Acceleration Framework in support to Member States, cooperation will be sought with other UN Agencies for complementary or joint training initiatives (with WMO for example on building capacity of national hydrological services) and other relevant partners including water-related NGOs and professional associations and private sector.

PRIORITY AREA 3 : BRIDGING THE DATA–KNOWLEDGE GAP

139 Transparency and accessibility of data are among the main pillars that sustain the advancement of open science — a coming commitment of UNESCO. Hydrological measurements are essential for decision-making and sustainable water resources management. The absence or inaccessibility of comprehensive or long-term data about water quantity, quality, distribution, access, risks, use, etc. often leads to partial or ineffective management and investments. Therefore, both sufficient data and its accessibility need to be ensured and, in many cases, improved.

140 Water data comes from various sources and data generators have a comparable diversity. However, the difficulty in collecting and understanding raw data and then applying it to a hydrological system in a decision context is often much more complex than initially contemplated. The gap between data and knowledge can only be bridged if data is collected in a transparent, comprehensible manner and can be scaled to the level of detail necessary to address specific issues. Appropriate data networks among riparian countries need to be established that enable data access from transboundary sources to all interested users. The challenge of data gathering, sharing, and interpretation becomes more complex when a water resource is transboundary. Therefore, there is a need to go beyond promoting data collection, to ensure data quality and to fill local gaps and add value to information.

141 Bridging the data-knowledge gap asks for availability, intergration and processing of data from various disciplines and sources; IHP-IX will take a full use of data and knowledge repositories available within the UNESCO water family, such as a Global Groundwater Information System (GGIS, since 2004) developed at the International Groundwater Resources Assessment Centre (IGRAC), a Category 2 UNESCO centre.

142 到 2029 年，水数据将更具透明性、可比较性和可获取性，可用于进一步开发开放型科学平台，为涵盖跨界水资源在内的综合流域水资源管理提供便利工具。

1 优先领域 3 与《2030 年可持续发展议程》之间的关系

143 水资源综合管理以及任何类型的科学和政策制定都必定要求充足可靠的数据条件。因此，改善数据的可获取性间接地支持了所有涉水可持续发展目标。具体来说，将通过监控与可持续发展目标 6 子目标相关的各种指标，支持相关行动决策以加速实施这些指标。只有通过加强伙伴关系（可持续发展目标 17）并根据可持续发展目标 6 的需求酌情开展跨界合作，才能使目标受众更好地获取和理解科学数据。

144 为了实现开放型科学愿景，政府间水文计划第九阶段战略规划将根据成员国要求，促进和推动以下与数据有关的成果。

2 预期成果

3.1 支持科学界开发和使用科学研究方法，以便正确收集、分析、解释和交换数据。

145 可靠数据是水资源管理的最重要基础，缺少可靠数据将严重阻碍决策实施。所有分析和建模都依赖于数据的数量、质量、覆盖度和可获取性。应扭转目前监测站点数量和采样频率下降的局面，保障数据的数量。数据质量决定了科学研究成果的质量。通过使用多样化的数据来源，提高了基于更大、更完整数据集的科学研究成果的可信度。应该尽可能将科学信息与本土 / 地方知识相结合。

146 然而，水数据不应局限于水量和水质参数。相反，还应监测用水趋势以及人类与地表水和地下水的其他相互作用。此外，元数据对数据验证至关重要，应该成为数据库的组成部分。

147 通过广泛合作，政府间水文计划第九阶段战略规划旨在提高水数据的数量、质量和有效性，在促进免费获取跨界水资源等所有水资源数据的同时，也将促进关于数据收集和交换战略以及分析方法的经验交流。它将进一步创建新能力 / 加强现有能力，促进各级合作。预计将与世界气象组织内的相关系统，如水文观测系统、世界水文循环观测系统、全球综合观测系统和信息系统等，建立紧密联系。

3.2 支持成员国、科学界和研究界建立统一的实验流域，以便收集科学数据，获取水文研究和全面水资源管理的知识。

148 在不同环境（三角洲、湿地、干旱地带、热带、小岛屿发展中国家等）背景下理解并考虑由于社会影响、气候变化等造成的水文循环变化是十分有必要的。小流域野外实验水文研究仍然是开发水文知识以及流域水文、气象和生物化学过程计算和预报方法不可或缺的数据来源。此外，这种尺度的研究也有助于监测由于自然和人为变化而造成的水文气象特征和气候变化机制状况的变化。

149 因此，应在联合国教科文组织水事机构的支持下管理和研究一系列实验流域，使其成为世界知识创造中心。可在这些流域开发和测试各种方法，收集有关可持续管理的科学信息。基于现有计划（如环境、生命和政策水文学计划），将尽可能地在诸如世界遗产地、生物圈保护区和全球地质公园等联合国教科文组织指定地点选择这类流域。

142 By the year 2029, significant advances will have occurred in transparency, comparability and accessibility of water data, which made possible further development of open-access science platforms and generated facilitating instruments for integrated watershed management, for all water resources, including transboundary ones.

Relation between Priority Area 3 and Agenda 2030

143 Sufficient and reliable data is an absolute must for IWRM and in general for any kind of science and policymaking. Improving the accessibility of data therefore indirectly supports all water-related SDGs. Specifically, all the SDG 6 targets will be reached through monitoring various related indicators and will support decision-making on actions to be taken to accelerate implementation. The better availability and comprehensibility of scientific data to the target audiences can only be achieved through strengthening partnerships (SDG 17) and through transboundary cooperation, as appropriate (SDG 6).

144 To realize the vision of open-access science, IHP-IX will be promoting and contributing to, among others, the following data-related outputs in line with the request of its Member States.

Expected outputs:

3.1 Development and use of scientific research methods by the scientific community supported to correctly collect, analyse, interpret and exchange data.

145 Reliable data is the most important basis for water resources management, without which implementation of decisions is severely handicapped. All analysis and modelling efforts are dependent on the quantity, quality, coverage, and accessibility of data. Data quantity should be maintained by reversing the current decline in the numbers of monitoring stations and sampling frequencies. Data quality determines the quality of scientific research outputs. The diversification of data sources allows scientific research to be based on larger and more complete data sets, increasing the confidence-levels of results. Scientific information should be combined with indigenous/local knowledge if available.

146 However, water data should not be limited to water quantity and quality parameters. Rather, water use trends and other human interactions with surface and groundwater should be monitored as well. Additionally, metadata are essential for data validation and should form an integral part of databases.

147 IHP-IX aims at improving the quantity, quality, and validation of water data in a broad collaborative effort and will promote the exchange of experiences in data collection and exchange strategies and analytical methodologies along with free access to data for all water resources, including transboundary ones. It will further create new/strengthen existing capacities and collaboration at all levels. Strong link to the WMO systems, including WMO Hydrological Observing System (WHOS), World Hydrological Cycle Observing System (WHYCOS), WMO Integrated Global Observing System (WIGOS) and WMO Information System (WIS) is envisaged.

3.2 Establishment of harmonized experimental basins by Member States, scientific and research communities, supported to collect scientific data and gain knowledge for hydrological research and holistic water management.

148 There is a need to understand and incorporate the changes to the hydrological cycle (such as social influences, climate change or others) in different environmental settings (delta, wetlands, arid, tropical, Small Island Developing States——SIDS, etc.). Experimental field hydrological studies in small catchments remain an indispensable source for the development of hydrological knowledge and methods for calculating and forecasting hydrological, meteorological and biochemical processes in river catchments. Additionally, monitoring of natural and anthropogenic changes in hydro-meteorological characteristics and regimes including climate change benefit from studies at this scale.

149 Thus, a chain of experimental basins should be managed and researched with the support of the UNESCO Water Family as hubs of knowledge creation all over the world. In these basins, methodologies can be developed and tested, and scientific information can be gathered on sustainable management. The basins will be selected based on existing initiatives like HELP and to the extent possible within UNESCO-designated sites, like World Heritage Sites, Biosphere Reserves and Global Geo-parks.

150 联合国教科文组织水事机构将协助建设全球实验流域网络。

3.3 支持科学界比较、验证并分享关于水量、水质和用水的开放性数据，以便促进可持续水资源管理。

151 拥有适当规模的、可用于制定规划和管理过程的数据，对于开发更好的决策支持系统、改善水治理、推进水教育，以及最终实现水资源可持续管理至关重要。除了可用性外，数据的准确性、可信度以及易于获取、便于理解的格式也是必不可少的。因此，数据维护应该是一个持续的过程。

152 获得可靠的、一致的数据对于开展包括跨界水资源在内的所有水资源数据比较研究和决策至关重要。专业人员需要为他们的研究计划获取必要数据，这些计划包括验证收集的数据，既从科学角度比较这些数据，也从政策角度理解如何应用这些数据。无论这些数据是通过传统野外技术还是通过最先进技术收集，无论是本地数据还是更大规模的数据，都需要尽可能地便于被获取。因此，将鼓励不同数据利益相关方共享数据。为了促进数据访问，在适当情况下应强调连接基于云端的数据库等现有网络数据库，开发数据访问应用编程接口，消除连接技术障碍（如平台之间的不兼容），以及完善基于网络的平台和质量保证协议。

153 除了基于最先进技术收集数据和通过传统方式交换协议外，还鼓励通过遥感、物联网采集、物联网传感器、私营部门和公民科学等收集和交换数据。

154 然而，由于投入精力有限和收集数据的不兼容性，多数情况下公民科学倡议并没有充分发挥其潜力。解决这类问题将会在更大范围内带来更好的科学和更健全的政策。为了能够准确解释公民科学数据，需要开发用户友好型平台、外联协议和能力建设活动，从而更好地让非政府组织和有关公民了解如何更有效地与决策者接触。

155 历史数据是理解趋势和罕见（极端）事件的基础。成员国和国际组织应该对历史数据、报告、会议记录和其他文件进行收集和数字化处理，并公布在网上，便于人们对这类事件的广泛理解。

156 上市公司和私营公司也在为各种目标收集现有水基础设施的运行数据。应根据成员国现行的“开放性访问”政策，最好在不同管辖范围内的公开数据库中公布这些数据。

157 政府间水文计划第九阶段战略规划鼓励引入联合国教科文组织其他部门的多学科数据，通过与水资源综合管理（社会、经济、环境）有关的自然科学、社会科学与水文学的结合，突出人类世时代对水资源的影响。政府间水文计划第九阶段战略规划将通过科学计划鼓励开放数据，支持成员国能力建设，以便开发包括针对公民和非政府组织平台在内的相关工具来访问数据，实现数据的可视化和连接。

3.4 加强科学界开发、分享和应用数据处理科学工具的能力（如数据同化和可视化方法，连接现有数据库和外联协议的质保协议）。

158 联合国水机制可持续发展目标 6 综合报告提出，需要创新科学方法，促进通过远程技术和公民科学获取的数据使用。还需要开发数据处理的新科学方法，并利用其他领域的前沿技术，帮助实现可持续发展目标等。人工智能和大数据技术将在这一过程中发挥关键作用。

159 为了正确收集、分析和解释现有数据，需要彻底理解并运用建模、预测、数据同化和数据可视化等科

150 UNESCO Water Family will assist the establishment of a network of experimental basins all over the world.

3.3 Comparing and validating open access data on water quantity, quality and use and their sharing by the scientific community supported for sustainable water management.

151 Availability of data at an appropriate scale, to undertake planning and management processes, is crucial to develop better decision support systems, improve water governance, advance water education, and eventually attain sustainable management of water resources. Besides availability, accuracy, credibility and an easily accessible and comprehensible format are also fundamental. Therefore, its maintenance should be a continuous process.

152 Access to reliable and consistent data is fundamentally important for comparative research and decision-making in all water resources, including in transboundary ones. Professionals need to be able to access necessary data for their purposes, including validating the data collected for comparison reasons both scientifically and for understanding how such data can be applied in a policy context. Whether this data is collected using traditional field techniques or using state-of-the-art technology, whether it is local data or data on a larger scale, it needs to be as accessible as possible. Therefore, data sharing of different data stakeholders will be encouraged. To facilitate data access, where applicable, emphasis should be put on connecting existing web databases, including cloud-based ones, developing data access Application Programming Interfaces (API) and removing technical obstacles to connectivity (e.g. incompatibilities between platforms), as well as in improving web-based platforms and quality assurance protocols.

153 Besides state-of-the-art data collection and exchange protocols by traditional means, remote sensing, Internet of Things (IoT) collection, IoT sensors, the private sector and citizen science can also be encouraged.

154 However, citizen science initiatives often times do not realize their full potential because of the limited reach of their efforts and the compatibility of the data collected. Solving these issues would lead to better science and sound policies on a larger scale. To enable accurate interpretation of citizen science data, user-friendly platforms, outreach protocols, and capacity building exercises need to be developed to better inform NGOs and concerned citizens as to how to engage with decision makers more effectively.

155 Historical data form the basis for understanding trends and rare (extreme) events. Member States and international organizations should collect, digitize and make available on the web historical data, reports, proceedings, and other documentation that will lead to a broader understanding of such events.

156 Public and private companies are also collecting data on the operation of the existing water infrastructure for various objectives. These data should ideally be posted in a publicly accessible database, at various jurisdictional scales, according to prevailing "open access" policies of Member States.

157 IHP-IX encourages the inclusion of multidisciplinary data from other UNESCO divisions and sectors to combine natural and social sciences related to Integrated Water Resources Management (social, economic, environmental) with hydrology to highlight influences on water resources in the Anthropocene era. IHP-IX will encourage open access data and support the capacity building of Member States in the development of tools for data accessibility, visibility and connectivity, including platforms targeting citizens and NGOs through scientific programmes.

3.4 Capacity of scientific community strengthened to develop, share and apply scientific tools for data processing (like data assimilation and visualization methods, quality assurance protocols to connect existing databases and outreach protocols).

158 The UN-Water SDG-6 Synthesis Report suggests the need for innovative scientific methods to enable the use of data from remote technologies and citizen science. Developing new scientific methods to process data and utilizing cutting-edge technologies from other sectors are also needed to help serve the SDGs and beyond. Artificial intelligence and big-data technologies will play a key role in this process.

159 To correctly collect, analyse, and interpret available data, scientific concepts like modelling, forecasting, data assimilation, and data visualization need to be thoroughly understood and practiced. Selection and correct use of any methodology is essential to be able to interpret the data in a way that is understandable by the broad scientific community. Additionally, creating capacities for better understanding by citizens, professionals and political authorities is vital to plan and implement water projects and to contribute to the achievement of water

学概念。为了能够使广大科学界以他们理解的方式解释数据，选择和正确使用何种方法至关重要。此外，培养公民、专业人员和政治当局更好的理解能力，对于规划和实施水项目以及实现水安全至关重要。为了便于决策和政策，将科学数据转化为可理解格式的方法通常是有限的。因此，有必要开发新思路并通过多种媒体传播新方法。

160 政府间水文计划将在第九阶段促进科学界开发和传播有关数据同化和可视化工具、质保协议、数据库链接和外联协议等方面的知识。政府间水文计划第九阶段战略规划将强化把数据映射为更好科学信息的科学研究方法，并通过分享新方法和工具，将科学数据转化为便于决策和政策制定的格式，扩大科学与政策之间的联系。

与其他联合国机构和合作伙伴的合作

161 为了更好地管理水资源，政府间水文计划将在第九阶段与联合国各机构和合作伙伴合作，致力于弥合数据与知识之间的鸿沟。隶属于联合国教科文组织水信息网络系统的教科文组织水计划 / 倡议正在并将继续为联合国水事机构数据平台和由联合国不同组织和计划主办的各种全球水数据中心做出重要贡献。这些数据平台有：联合国环境规划署监测淡水的全球淡水环境监测系统、世界气象组织水文学和水资源计划、水文观测系统、世界水文循环观测系统、全球水文测验支持设施等。数据中心有：世界气象组织全球降水气候学中心、全球径流数据中心、全球冰冻圈观察、国际湖泊和水库水文学数据中心、国际土壤水分网络、联合国粮农组织全球水与农业信息系统和水资源生产力开放门户、联合国环境规划署全球淡水环境监测系统数据中心、世界冰川监测服务等。它们全部与联合国教科文组织二类中心国际水资源和全球变化中心下属的全球陆地水文学网络组成联盟。联合国教科文组织将与世界气象组织密切合作，为水和气候联盟做出贡献，继而为联合国可持续发展目标 6“全球加速器框架”提供数据和信息。联合国教科文组织二类中心——国际地下水资源评估中心与世界气象组织全球地下水监测网络（创建于 2007 年）的出色合作是值得学习的典范。联合国教科文组织还将在“综合监测倡议框架”下与负责监管可持续发展目标 6 不同指标的联合国机构继续合作，分享与可持续发展目标 6 监测有关的数据。

四、优先领域 4：全球变化条件下的水资源综合管理

162 健康的河流、湖泊、湿地、含水层和冰川不仅提供安全饮用水，保障生物多样性，维护地球上所有生态系统；而且支持农业、水电、工业、娱乐、通信和货物运输。虽然水被认为是可持续社会经济发展的核心，但在投资辩论中经常被忽视。此外，目前没有以综合的方式开展水管理，水管理被认为是许多不同政府机构共同的责任。

163 全球变化对水资源综合管理既是威胁也是机遇。水管理应该具有包容性，以综合的视角，采用关系方法和从源头到海洋的方法，加强各种机制，促使所有水利益相关方参与。这也意味着实现水安全的同时，要保护水质、保障环境流量以及包括淡水供应在内的生态系统服务，免受各种来源、各种利益、各级政府和尽可能广泛的相关学科影响。

164 到 2029 年，社会上大多数人已设法适应或缓解了由气候变化和诸如全球流行病等人为因素引起的水风险，为地球的未来创造更好的参与式管理实践和新机遇。

security. The current methods for translating scientific data into an intelligible format for decision-making and policy formulation are, in general, limited. Therefore, it is necessary to develop new ideas and disseminate new methods through multiple media.

160 IHP will work during its ninth phase as a catalyst for development and dissemination of knowledge about data assimilation and visualization tools, quality assurance protocols, database links and outreach protocols within the scientific community. IHP-IX will enhance scientific research methods to map data into better scientific information and widen the science-policy interface by sharing new methods and tools translating scientific data into a format facilitating decision-making and policy formulation.

Cooperating with other UN Agencies and partners

161 IHP in its ninth phase will cooperate with various UN agencies and partners in contributing to bridging the data-knowledge gap for better water management. UNESCO water programmes/initiatives connected under the UNESCO Water Information Network Systems (WINS) are and will be making essential contributions to the data platforms of the UN Water Family like the Global Environment Monitoring System for freshwater (GEMS/Water by UNEP), the Hydrology and Water Resources Programme (HWRP by WMO), the WMO Hydrological Observing System (WHOS), the World Hydrological Cycle Observing System (WHYCOS), the Global Hydrometry Support Facility (WMO HydroHub) and various global water data centres under auspices of different UN organisations and various global water data centres under auspices of different UN organisations and programmes such as the Global Precipitation Climatology Centre (GPCC/WMO), Global Runoff Data Centre (GRDC/WMO), International Data Centre on Hydrology of Lakes and Reservoirs (HYDROLARE), International Soil Moisture Network (ISMN), AQUASTAT and WaPOR (FAO), GWDC (GEMS/Water Data Centre/UNEP), Global Cryosphere Watch (GCW/WMO), the World Glacier Monitoring Service (WGMS), all federated in the Global Terrestrial Network Hydrology (GT-NH) hosted at the International Centre of Water Resources and Global Change (ICWRGC), a UNESCO Category 2 Centre (C2C). UNESCO will closely cooperate with WMO in contributing to the Water and Climate coalition as contribution to the data and information component of the UN SDG 6 global acceleration framework. The excellent cooperation with WMO on the Global Groundwater Monitoring Network (GGMN, since 2007) implemented by UNESCO's C2C, IGRAC, will be used as an example to follow. UNESCO will also continue to cooperate with UN agencies custodians of the different SDG 6 indicators under the Integrated Monitoring Initiatives (IMI) for sharing data related SDG 6 monitoring.

PRIORITY AREA 4 : INTEGRATED WATER RESOURCES MANAGEMENT UNDER CONDITIONS OF GLOBAL CHANGE

162 Healthy rivers, lakes, wetlands, aquifers, and glaciers do not just supply safe drinking water, safeguard biodiversity and maintain all ecosystems on the planet; they also support agriculture, hydropower, industry, recreation, communications, and transportation of goods. Although water is considered the core of sustainable socio-economic development, it is frequently ignored in the investment debate. Additionally, water management is not considered in an integrated manner and is frequently considered a shared responsibility among many different governmental institutions.

163 Global change is simultaneously a threat and an opportunity for integrated water management. Water management should be inclusive to strengthen all the mechanisms that enable the participation of all water stakeholders, with an integrative perspective using the nexus and source-to-sea approaches. It also means achieving water security while protecting water quality, the environmental flows and its ecosystems services, including all fresh water, independent of its diverse sources, all interests, all levels of government, and the widest possible range of relevant disciplines.

164 By 2029, most societies have managed to adapt to or mitigate water risks derived from, among others, climate change and the human factor, such as global pandemics, generating better participatory management practices and new opportunities for the future of our planet.

1 优先领域 4 与《2030 年可持续发展议程》之间的关系

165 该优先领域的发展和实施补充了《2030 年可持续发展议程》中的有关目标，即可持续发展目标 6.1 和 6.2（普遍和公平获得饮用水、环境卫生和个人卫生）、可持续发展目标 6.3（改善水质）和可持续发展目标 6.5.1（实施各级水资源综合管理），包括了可持续发展目标 6.5.2（酌情跨界合作）、可持续发展目标 6.6 和 15.1（保护和恢复山区、森林、湿地、河流、含水层和湖泊等与水有关的生态系统）。

166 其他目标与下列目标有关：防治荒漠化、恢复受干旱和洪水影响而退化的土地和土壤、可持续发展目标 15.3（实现土地退化中性的世界）、可持续发展目标 11.B（大幅减少水灾致死和受影响的人数）、可持续发展目标 6.B（加强当地社区参与改善对水和卫生设施的管理）以及加强可持续发展全球伙伴关系。通过多利益相关方伙伴关系调动和分享知识、专长、技术和财政资源，实现所有这些目标的互补，以便支持可持续发展目标 17（实现所有国家的可持续发展）。

2 预期成果

4.1 支持科学界开展并分享关于包容性和参与性方法的研究，确保通过支持青年、地方和土著社区以公开、积极、有意义的、促进性别平等的方式，带动所有利益相关方参与到水资源管理过程中。

167 如果在咨询和决策过程中一直不采用综合和包容的方式考虑水问题，那么水政策和水资源综合管理之间的差距就会变大。

168 社会弱势和少数群体（尤其是妇女、青年、土著群体、少数民族）适当参与水资源管理，可以强化水资源问责制和责任制，形成负责且包容的水资源管理。参与的核心要点是在所有决策过程中实现性别平等。参与式管理还包括促进公民科学、以用户为中心的设计、青年和社区参与。还应特别关注实际水资源用户群体，并最终由他们实施水资源管理决策和实践。许多与水有关的不平等现象是由权力关系造成的。通过以伙伴关系形式共同努力，可以对水资源管理产生更广泛、更被认可的影响。

169 为了解决无效和低效的水资源管理实践问题，政府间水文计划将强化参与性管理方法，促进践行包容性水资源管理理念，并确保以年轻人的思想以及本土和地方知识为出发点，让所有利益相关方参与到水资源管理过程中。将加强专家按性别分列数据的能力，促进性别平等的水资源管理。

4.2 科学界和联合国教科文组织水事机构开展并分享有关河流上下游水力发电、航运、渔业、休闲活动、供水、洪旱风险管理的研究，以便尽量降低对社会、经济和生态的负面影响。

170 很显然，成员国需要加强对水循环管理方法的研究，以便实现流域级的全面水资源管理。在应用水循环管理方法时，成员国可以采用可持续发展方式，同时满足人类和环境目标，实现水安全目标。为了有效实施水循环管理方法，必须建立一个能够评估流域内水文循环过程的系统。此外，还需要采用河流综合研究与管理方法，改善上下游对河流的使用，尽量减少全球变化对河流的影响，并改善河流的生态系统服务。

171 反映水资源管理的整体方法是至关重要的，即考虑能源（水电）、交通（水运）或洪水风险管理对上下游河流使用、社会经济及生态的综合影响。这一方法特别有助于解决跨界水问题。考虑到河流使用对水、泥沙和生态连续性的影响，这种上下游一体的方法既包括了影响研究，也包括了改善河流管理，

Relationship between Priority Area 4 and Agenda 2030

165 The development and implementation of this Priority Area complements those targets of the Agenda 2030 related to universal and equitable access to drinking water, sanitation and hygiene (6.1 and 6.2), improving water quality (6.3), implementing integrated water resources management at all levels (6.5.1), including transboundary cooperation as appropriate (6.5.2) and protecting and restoring water-related ecosystems, including mountains, forests, wetlands, rivers, aquifers and lakes (6.6 and 15.1).

166 Other targets are related to combating desertification, restoring degraded land and soil, including land affected by drought and floods, and achieving a land degradation-neutral world (15.3), significantly reducing the number of deaths and of people affected by water-related disasters (11.B) while strengthening the participation of local communities in improving water and sanitation management (6.B), and enhancing the global partnership for sustainable development. All these targets are complemented by multi-stakeholder partnerships that mobilize and share knowledge, expertise, technology, and financial resources, to support the achievement of the SDGs in all countries (SDG 17).

Expected outputs:

4.1 Conducting and sharing of research on inclusive and participatory approaches by the scientific community, to ensure open, active, meaningful gender-responsive engagement of youth, local and indigenous communities supported to enable all stakeholders to be part of the water management process.

167 The gap between water policies and its integrated management can become great, if it is not always considered in an integrated and inclusive manner, including in consultative and decision-making processes.

168 Social participation of vulnerable and minority groups, (inter alia of women, youth, indigenous groups, national minorities) may lead to improving water accountability and responsibility and, when properly achieved, can lead to conscientious and inclusive resource management. A central component of participation is gender equality in all decision-making instances. Participatory management also includes the enabling of citizen science, user-centred design, youth and community participation. Special attention should also be given to the actual water resource user groups that in the end have to deal with the resulting water management decisions and practices. Many water-related inequalities are due to power-relations. By working together in partnerships, more wide-ranging and accepted impacts can be achieved.

169 To tackle ineffective and inefficient water management practices, IHP will work towards enhancing participatory management methodologies promoting the concept that water management efforts should be implemented through inclusive approaches and ensuring that young minds, indigenous and local knowledge are the starting point, having ensured all stakeholders are included in the process. Capacity of experts will be strengthened on the gender-disaggregated data for gendered water management.

4.2 Research on upstream-downstream river uses for hydropower, navigation, fishery, leisure activities, water supply, drought risk management and flood risk management conducted and shared by the scientific community and UNESCO Water Family to minimize socio-economic and ecological consequences.

170 The need to enhance research on Water Cycle Management (WCM) methods by Member States is obvious if they are to implement holistic management of their water resources at the watershed level. When applying WCM methods, one can satisfy both human and environmental objectives in a sustainable manner while aiming towards water security. In order to effectively implement WCM, a system that can evaluate the hydrological circulation process in the watershed must be established. Furthermore, an Integrated River Research and Management (IRM) is needed to improve upstream-downstream river uses, minimize Global Change effects and improve ecosystem services of rivers.

171 A holistic approach reflecting water management from the perspective of the upstream-downstream integration of river usages and socio-economic and ecological consequences with respect to energy (hydropower), transport (navigation) or flood risk management is crucial. This is particularly helpful in addressing transboundary water issues. The upstream-downstream integration includes the research and improved river management concerning water, sediment and ecological continuity being affected by river uses and better understanding trajectories of change over time.

以便深入了解这些影响随时间的变化情况。

172 政府间水文计划第九阶段战略规划将就水循环管理方法、河流综合研究与管理开展能力建设，支持基础研究，并利用现有政府间水文计划倡议确定最佳实践，以便向更广大社区提供科学、合理且实用的数据。

4.3 支持科学界开展并分享关于废水回用、海水淡化、雨水收集和含水层补给管理等非常规水资源的研究，以便改善水循环管理，加强地方、区域和国家决策者的能力，提高公众认可度。

173 非常规水资源是水循环管理的重要基础，它包括废水回用、海水淡化、雨水收集和雾气收集。非常规水资源最广泛的用途是将处理过的废水用于农业灌溉。经处理和未经处理的废水是一种替代性的非常规水资源，可以被安全利用，在降低水污染的同时允许回收有用的营养物和能源等副产品。然而，有必要提升污染物对健康和环境风险影响的认识和管理实践，确保安全利用再生水资源。海水淡化为那些极度缺水且能大量获取海水的国家提供了源源不断的水源。但是，由于海水淡化需要能源且会产生对环境有负面影响的物质（如浓盐水），所以在科学和创新发展中应考虑生态系统 / 环境问题、能源问题和经济问题。

174 政府间水文计划将继续努力促进实施水循环管理，在强调应用低成本新技术的前提下，根据各地实际情况合作开发、使用非常规水资源。政府间水文计划将探索如何在规划可持续用水需求时考虑非常规水资源。将在南北半球确定和分享关于废水回收、废水处理、含水层补给管理和脱盐技术等的最佳实践。它还将进一步提高公众对非常规水资源的认识，通过专家培训来提高公众对使用非常规水资源的接受程度。

4.4 支持科学界开发和分享有关利用“源头到海洋”和关系方法的知识，以便加强能力建设、改善包括跨界流域在内的所有流域的水资源综合管理。

175 水从最高山脉的源头穿过流域和含水层流向海洋。“源头到海洋”的方法有助于更好地管理近海景观、减少洪水风险、保护地下水补给区、维持生态系统和河口区的健康。此外，这种方法提高了人们对河流流量、土壤水和地下水之间关系的理解。当地表水源枯竭或受到严重污染而无法通过较为经济的方式进行修复时，理解它们之间的关系尤为重要。

176 政府间水文计划第九阶段战略规划将为全面、科学地理解“源头到海洋”现象提供知识基础，促进生成有关源头到海洋相互联系特别是与水资源有关的全球知识，并向成员国提出适应全球变化的方案。通过直接与联合国教科文组织政府间海洋学委员会、其他伙伴的国际源头到海洋平台开展合作，政府间水文计划将进一步促进和支持使用“源头到海洋”方法开展案例等其他研究。

177 不应把水、粮食、能源和生态系统看作是独立的存在，而应把它们看作是复杂且不可分割的部分，需要解决其相互关联的资源、挑战和响应措施。关系方法则提供了一种涵盖所有部门、考虑用户需求且应被进一步应用到日常水管理实践中的方法。但仍需要通过探索关系方法的科学、研究和创新，确定相互依存部门之间的协同作用、权衡关系，应对复杂的全球发展和安全挑战，支持在流域到全球所有尺度上实施可持续发展目标。

178 为了应对这些挑战，政府间水文计划第九阶段战略规划将支持一种综合方法，这种方法具体阐述涉水部门和其他部门之间的紧密联系，例如粮食和能源之间的关系，鉴于这些联系在提高效率、减少权衡

172 IHP-IX will build capacity on the WCM approach and IRM and support fundamental research and identification of best practices using existing IHP initiatives with the aim of providing scientifically sound and usable data to the larger community.

4.3 Conducting and sharing of research on Non-Conventional Water Resources (NCWRs) such as wastewater reuse, desalination, rainwater harvesting, and the Management of Aquifer Recharge (MAR) by the scientific community, in support of improving Water Cycle Management (WCM), strengthening capacities of local, regional, and national decision-makers, and enhanced acceptance of public.

173 Non-Conventional Water Resources (NCWRs) are an important underpinning of WCM, including such as wastewater reuse, desalination, rainwater harvesting and fog harvesting. The most widespread use of NCWRs is using treated wastewater for agricultural irrigation. The safe and beneficial use of treated and untreated wastewater offers an alternative non-conventional water resource, while reducing water pollution and allowing for the recovery of useful by-products such as nutrients and energy. Yet, there is a need to improve knowledge and management practices to ensure safe water reuse, in particular regarding health and environmental risks of pollutants. Desalination provides a constant source of water in countries that face extreme scarcity and have access to the most abundant form of water found in seas and oceans. However, the 3-E issues (Ecosystems/ Environment, Energy and Economics) should be taken into consideration in scientific and innovation development, as desalination requires energy and produces secondary effects with negative impacts on the environment (brines for example).

174 IHP will continue its efforts to promote better implementation of the WCM by collaborating in it the development and implementation of NCWRs based on different regional realities with a particular emphasis on new technologies with lower costs. IHP will explore how NCWR can be incorporated when planning sustainable use of water demand. Best practices will be identified and shared between the Global North and Global South concerning for example recycling, wastewater treatment, MAR, and desalination technologies. It will further raise awareness and train experts to enhance acceptance of the general public in using such resources.

4.4 Development and sharing of knowledge on using the source-to-sea and nexus approaches by the scientific community supported, and capacities strengthened to improve integrated water resources management for all watersheds, including transboundary ones.

175 Water flows from sources in the highest mountains to the sea or ocean through river basins and aquifers. The source-to-sea approach contributes to better management of near-shore landscapes, reduces flood risk, allows for protection of groundwater recharge zones, and maintains healthy ecosystems as well as estuarine zones. Additionally, this approach improves understanding of the relationships between river flow, soil water and groundwater, which is becoming more important as surface water sources dry up or become too polluted to economically clean up.

176 IHP-IX will provide the knowledge base to develop a comprehensive scientific understanding of the source-to-sea phenomena and will contribute to global knowledge generation on source-to-sea interconnections, particularly related to water resources and as well in proposing options for adaptation to the Member States. IHP will further promote and support additional research and case studies in using the source-to-sea approach, benefiting from direct cooperation with UNESCO's Intergovernmental Oceanographic Commission (IOC) and other partners of the international source-to-sea platform.

177 It is necessary that water, food, energy, and ecosystems be viewed not as separate entities, but rather as complex and inextricably intertwined sectors needed to address interconnected resources, challenges as well as responses. The use of the nexus approach provides an all sector encompassing way of considering user requirements and should be further implemented into daily water management practices. There is still much to learn, related to the science, research, and innovation part of the nexus approach, to identify synergies and trade-offs between interdependent sectors to address the complex global development and security challenges and support the implementation of the SDGs at all scales ranging from watersheds to a global scope.

178 To address these challenges IHP-IX will support an integrated approach that concretely addresses the most relevant interlinkages among sectors beyond and within water such as food and energy, considering that these interlinkages can increase efficiency, reduce trade-offs, and build synergies while improving governance across sectors.

和建立协同作用的同时能改善跨部门治理。

179 通过提供实施关系方法的证据、情景、具体工具，政府间水文计划第九阶段战略规划将支持基于关系方法的知情决策，并支持利益相关方参与到决策过程中。此外，通过与联合国粮农组织、欧盟联合研究中心等其他伙伴协调，政府间水文计划第九阶段战略规划将促进关系方法的使用，协调各部门和利益相关方，实现协同增效并管理经常性竞争的利益要素，同时确保生态系统的完整性。

4.5 科学界和联合国教科文组织水事机构提高对地表水（河流、湖泊、湿地）和地下水淡水系统的污染物来源、归宿和迁移的认识和了解，以便预防和降低水污染、支持水资源管理战略。

180 世界各地淡水资源污染不断加剧，呼吁采取紧急行动，降低淡水污染对人类健康和水生生态系统的负面影响。水污染也是导致生态系统退化和生物多样性丧失的主要直接因素之一。由于人类缺乏对诸如药品和微塑料等新兴和新污染物在健康和生态毒理学方面潜在风险的了解，需要特别关注此类污染物。

181 政府间水文计划第九阶段战略规划将重点加强对水生环境中各类型污染的科学认识和评估，为循证管理决策和基于科学的适当政策响应提供基本依据。将在污染物生命周期的各个阶段，采取提高人们对可持续生产和消费的认识、改善废水处理、再利用和管理等措施，预防、减少并控制向水生环境排放的各类污染物。新兴污染物（药物和化学品）在淡水环境中具有潜在的健康和生态风险，必须对它们进行毒性和环境影响研究，加强风险评估的科学基础，所以政府间水文计划第九阶段战略规划将继续开展相关研究。它将制定并推广适当的“末端”污染治理方案，包括低成本废水处理技术。它将寻求与联合国环境规划署等其他联合国机构的伙伴关系，实现互补和协同增效。

4.6 支持科学界对生态水文学试验点的生态系统服务和环境流量进行评估、分享评估成果，以便改善水资源综合管理。

182 成功的淡水资源管理需要对生态系统服务进行评价并加强其功能，需要特别强调环境流量，以便全面实施淡水资源管理方法。水生态系统可持续管理面临日益增长的挑战，生态水文学则提供了绿色解决方案，它结合了水文学、生物群和工程方法，通过改善水质、提高水量来实现水安全。在流域层面，通过整合并协调生态水文学解决方案、基于自然的解决方案以及水文技术基础设施，利用低成本提高措施效率，有效避免因水资源管理和水资源分配而造成的冲突。

183 通过联合国教科文组织政府间水文计划的生态水文网络，特别是 19 个国家的 26 个示范点，已经了解了水 – 生态系统 – 社会之间的相互作用，并已将这一成果转化成了创新生态水文学解决方案和基于自然的解决方案。这些示范点是培训专业人员和开展 / 参与此类社会教育的实验室，并由水、生物多样性、生态系统的社会服务、影响抵御力以及社会、文化和教育组成部分形成提高流域可持续性潜力的根本。

184 政府间水文计划旨在加强对水 – 植物 – 土壤 – 地下水相互作用等生态系统过程之间关系的理解，通过减少洪水风险、保护地下水补给区以及维持生态系统和河口区健康来更好地管理流域景观。

185 虽然能力建设不是新概念，但是政府间水文计划第九阶段战略规划将更多地支持能力建设活动，在水务部门筹划未来投资中使用水资源综合管理和生态水文学工具。

179 IHP-IX will support informed decision on nexus by providing evidence, scenarios, concrete tools to implement nexus approach and supporting stakeholder engagement. Furthermore, IHP-IX will promote, in coordination with other partners such as FAO and JRC, the nexus approach for coordination across sectors and stakeholders to enable synergies and managing often competing interests while also ensuring the integrity of ecosystems.

4.5 Understanding and knowledge on pollutants sources, fate and transport in freshwater systems, including surface waters (rivers, lakes, wetlands) and groundwater improved by the scientific community and UNESCO Water Family to prevent and reduce water pollution and underpin water resources management strategies.

180 Freshwater resources pollution is worsening throughout the world, calling for urgent action to reduce negative effects on human health and aquatic ecosystems. Water pollution is also one of the main direct factors that cause ecosystem degradation and biodiversity loss. Emerging and new pollutants such as pharmaceuticals and microplastics pose a particular concern due to the lack of knowledge on their potential health and ecotoxicological risks.

181 IHP-IX will focus on improving the scientific understanding and assessments of all types of pollutants in the aquatic environment providing thus a fundamental basis for evidence-based management decisions and appropriate science-based policy responses. Measures to prevent, reduce and control discharges of all types of pollutants to the aquatic environment will be promoted at all stages of the life cycle of pollutants—from awareness raising on sustainable production and consumption, to improving wastewater treatment, reuse, and management. Research on emerging pollutants (pharmaceuticals and chemicals), on their toxicity and environmental effects, is essential to enhance the scientific basis of risk assessment in relation to their potential health and ecological risks in freshwater environments and will be pursued. Appropriate 'end-of-the-pipe' pollution abatement solutions, including low-cost technologies for wastewater treatment, will be identified and promoted. Partnership with other UN agencies such as UNEP will be sought for complementarity and synergy.

4.6 Undertaking and sharing assessment of ecosystem services and environmental flows in ecohydrology pilot sites by the scientific community supported, to improve integrated water resources management.

182 Successful management of freshwater resources requires an assessment and enhancement of ecosystem services, with special emphasis on environmental flows to fully implement this approach. Ecohydrology creates green solutions for increasing challenges in the sustainable management of water ecosystems. It combines hydrology, biota and engineering for water security, to enhance both water quality and quantity. The integration and harmonisation of ecohydrological solutions and nature-based solutions with hydrotechnical infrastructure at catchment level, improves efficiency of measures at lower costs, thus helping to avoid conflicts for water resources management and water allocation.

183 The UNESCO-IHP ecohydrology network and especially the demonstration sites (26 in 19 countries) has developed the understanding of water-ecosystems-society interactions, translated into innovative ecohydrological solutions and nature-based solutions used as laboratory for training of professionals and for society education/involvement, are fundamental for enhancement of the catchment sustainability potential through WBSRCE (Water, Biodiversity, ecosystem Services for Society, Resilience to impacts, social, Cultural and Educational components).

184 IHP aims at improving the understanding of the relationships between ecosystem processes such as the water-plants-soil-groundwater interactions, to better manage catchment landscapes by reducing flood risks, allowing for protection of groundwater recharge areas, and maintaining healthy ecosystems as well as estuarine zones.

185 While not a new concept, IHP-IX will give more support to capacity building activities in the use of integrated water resource management and ecohydrology tools in organizing future investments in the water sector.

4.7 支持科学界评估、制定和分享用于监测冰冻圈系统（雪、冰川和永久冻土）、冰川融化形成的径流、泥沙侵蚀和输移、冰川水库（山地湖泊等）和含水层的变化的方法，以便理解它们的潜在用途，并为各级决策者提供信息。

186 在过去的几十年里，全球变暖导致了冰冻圈大范围萎缩、冰川大量消失、雪盖减少、永久冻土温度升高，给依赖冰冻圈水资源的社会带来了巨大的风险。同样，作为饮用水的主要来源，含水层开采量的不断上升也往往造成不可逆转的影响。

187 政府间水文计划将促进科学研究，加强对冰冻圈和含水层供水能力的了解，并呼吁科学决策。通过评估和促进区域合作以及利益相关方参与，政府间水文计划第九阶段战略规划将加强各国适应气候变化影响的能力。

188 政府间水文计划第九阶段战略规划将支持加强国家和地区对含水层和冰湖溃决洪水灾害的监测及评估能力。将对若干地区的可用地下水和冰冻圈的变化进行情景分析，以便应对气候变化。

4.8 支持科学界开发和分享将全球变化纳入水资源管理的方法和工具，以便完善各级决策者规划。

189 目前还没有达成关于全球变化对淡水资源管理影响的共识，因此水资源管理的理论和实践将不得不继续适应地球当前现状和未来趋势。为了进一步基于生态水文合理性开展水资源时空优化管理，需要既重视缓解全球变暖，又重视适应全球变暖和提高诸如小岛屿发展中国家、半干旱地区、湿地、沿海腹地和山区等风险敏感地区的韧性。应该大力援助这些地区开发并向其分享应对全球变暖的新方法和工具。

190 时空优化反映了“区域特定条件”，同时考虑了每个研究区季节性以及水力和水文因素。在实现时空优化的过程中，必须考虑生态和水文因素及其影响。仅仅注重人类利益的战略可能会破坏生态系统，从长远来看还会造成粮食和农业问题。

191 政府间水文计划第九阶段战略规划将更加重视开发和分享应对这些挑战的方法、指南和工具，特别是与环境敏感型自然资源有关的挑战。

4.9 支持成员国在联合国水机制和联合国欧洲经济委员会协调下，通过适当的跨界合作，实施各级水资源综合管理，以便实现可持续发展目标 6.5。

192 目前，已经确定了 260 多个跨界流域和 600 多个跨界含水层。跨界水资源需要由沿岸国家以互利方式进行管理。《2030 年可持续发展议程》中的可持续发展目标 6.5 强调了跨界合作的重要性。跨界合作监控能够促进成员国对沿岸国家合作状态的评价，并为改善协调方式而设定目标。作为可持续发展目标 6.5.2 指标的共同监管机构，联合国教科文组织将与联合国欧洲经济委员会一起继续支持成员国开展监测、评估和发展跨界合作，尤其在地下水资源及其管理方面，减小国家能力上的主要差距。这将有助于各国开展合作谈判。

3 与其他联合国机构和科学伙伴的合作

193 可持续发展目标 6.5.2 是《2030 年可持续发展议程》中唯一明确的跨界水合作目标；作为该目标指标的监管机构，联合国教科文组织与联合国欧洲经济委员会共同发挥着重要作用。

4.7 Undertaking assessments and developing and sharing of methods to monitor changes in the cryosphere system (snow, glacier, and permafrost), runoff formation from melting glaciers, erosion and sediment transport, glacier-fed reservoirs such as mountain lakes, and aquifers, by the scientific community supported for improved understanding of their potential use to inform decision makers at all levels.

186 Over the last decades, global warming has led to widespread shrinking of the cryosphere, with mass loss of glaciers and reductions in snow cover and increased permafrost temperature created profound risks for societies that depend on cryosphere for water resources. Similarly, aquifers, as the major sources of potable water, have been experiencing increased abstractions that often cause irreversible effects.

187 IHP will promote scientific research to enhance understanding of water availability from the cryosphere and in aquifers and call for science-based policy decisions. IHP-IX will strengthen the adaptation capacity of countries to climate change impacts through assessment, promotion of regional cooperation, and stakeholder engagement.

188 IHP-IX will support strengthening national and regional capacities to monitor and assess aquifers and Glacial Lake Outburst Flood (GLOF) hazards. Scenarios of changes in the available groundwater and cryosphere in response to climate change for several regions will be implemented.

4.8 Development and sharing of methodologies and tools in mainstreaming global changes within water management by the scientific community supported for improved planning by decision makers at all levels.

189 The impacts of global changes on the management of freshwater resources are hardly recognized, therefore, theory and practice of water resources management will have to continue to adapt to current and future trends facing the planet. Further to spatial-temporal optimization of water resources management based on ecohydrological soundness, it is necessary to both focus on mitigating global warming, as well as on adapting to it and increasing resilience in risk sensitive areas such as SIDS, semi-arid regions, wetlands, coastal hinterlands, and mountainous areas, which should receive major assistance in the development and sharing of new methodologies and tools.

190 The spatial-temporal optimization reflects the “regional-specific condition” considering the seasonality and the hydraulic and hydrologic factors of each study area. It is essential to consider the ecological and hydrological factors and effects in achieving the spatial-temporal optimization. The strategy focusing on the human benefit alone may destroy the ecosystem and cause a food and agricultural problem in the long-term perspective.

191 IHP-IX will give more emphasis to develop and share methodologies, guidelines, and tools to address these challenges, particularly related to environmental sensitive natural resources.

4.9 Implementing integrated water resources management at all levels, through transboundary cooperation as appropriate by Member States, supported, in coordination with UN-Water and UNECE, to achieve SDG target 6.5.

192 Currently, over 260 transboundary river basins and more than 600 transboundary aquifers have been identified. Transboundary water resources need to be managed in a mutually beneficial manner by all riparian states. Agenda 2030 through Target 6.5 emphasizes the importance of transboundary cooperation. Monitoring of transboundary cooperation provides impetus for countries to assess the status of cooperation with neighbouring countries and set targets for improved coordination. UNESCO, as the co-custodian agency of SDG 6.5.2 indicator, will continue, together with UNECE, to support Member States in monitoring and to assess and develop their transboundary cooperation, tackling the main gaps in the national capacity, especially regarding groundwater resources and their management. This will help countries to negotiate cooperative arrangements.

Cooperating with other UN Agencies and scientific partners

193 UNESCO has an important role, together with UNECE as custodian agencies for 6.5.2, the only target in the 2030 Agenda explicitly related to transboundary water cooperation.

194 为了促进水资源综合管理，联合国教科文组织将动员联合国环境规划署、联合国开发计划署、世界气象组织、联合国欧洲经济委员会和联合国粮农组织等不同联合国机构成员、学术与研究机构，以及全球水伙伴等非政府组织开展工作，形成互补。政府间水文计划第九阶段战略规划的主要贡献在于开展能力建设，并提供基于科学的知识、工具、方法和指南。

五、优先领域 5：基于科学的缓解性、适应性和韧性水治理

195 水治理是指现有政治、社会、经济、法律和行政体系，它们对取用水、防止水污染和水资源日常管理产生一定影响。它决定了水资源和服务分配的公平性和效率性，并通过保护生态系统使社会经济活动用水与提供产品和服务的生态系统用水之间达到平衡。水治理包括制定、建立和实施水伦理、法律和制度等水政策，基于科学的、明确实用的标准，以及所有利益相关方角色和职责。例如，联合国大会关于跨界含水层法的第 A/RES/68/118 号决议，说明了如何通过联合国教科文组织的科学支持助力与水治理有关的准备工作。该决议呼吁并鼓励联合国教科文组织政府间水文计划继续向有关国家提供进一步的科技援助。

196 到 2029 年，通过利用基于科学的工具、能力和知识，成员国能够解决适应和缓解气候变化方面的问题，显著缩小水治理差距。

1 优先领域 5 与《2030 年可持续发展议程》之间的关系

197 良好水治理是制定可持续发展目标等全球整体目标的基础，因此它与可持续发展目标 6（确保所有人对水和卫生设施的使用和可持续管理）的子目标 6.4（用水效率）、子目标 6.5.1（水资源综合管理）、子目标 6.5.2（酌情跨界合作）、子目标 6.A（支持国际合作和水能力建设）、子目标 6.B（地方社区参与到决策过程中）直接相关。治理进展也影响可持续发展目标 1（消除贫困）和可持续发展目标 2（消除饥饿）、以及可持续发展目标 1.4 和 1.5（增强抵御能力并降低暴露在极端天气事件中的机会）、可持续发展目标 2.3（使小规模粮食生产者生产力和收入翻倍）、可持续发展目标 2.4（执行具有抗灾能力的农作方法，加强适应气候变化能力）。

198 水治理与可持续发展目标 3（良好的健康和福祉）之间存在重要联系，尤其与子目标 3.9（减少由于危险化学品以及空气、水和土壤污染造成死亡和患病人数）关系密切。水治理与可持续发展目标 4（优质教育）之间也有重要联系，尤其与子目标 4.1 和 4.5（消除教育中的性别差距和所有歧视）关系密切。同样，水治理与可持续发展目标 5（性别平等）之间的联系强调了子目标 5.1（消灭对妇女和女童一切形式的歧视）和子目标 5.B（加强技术应用，促进妇女赋权）。水治理还与可持续发展目标 8（包容性和可持续经济增长、就业和人人都有体面工作）之间存在联系，体现在子目标 8.2、8.3、8.4 和 8.9（通过多样化经营、技术升级和创新实现更高水平的经济生产力，并努力使经济增长与环境退化脱钩）上。

199 本优先领域将推动实现可持续发展目标 10（减少国家内部和国家之间的不平等）、可持续发展目标 11（建立包容、安全、韧性和可持续的城市）、可持续发展目标 13（采取行动应对气候变化及其影响）及其子目标 13.1（加强各国抵御和适应气候相关的灾害和自然灾害的能力）、13.2（将应对气候变化的举措纳入国家政策、战略和规划）和 13.B（促进最不发达国家和小岛屿发展中国家建立增强能力的机制，帮助其进行与气候变化有关的有效规划和管理，包括重点关注妇女、青年、地方社区和边缘化社区）。健全的水治理是可持续发展目标 16（和平、正义和强大机构）的基础，它也与可持续发展子目标 6.5（酌情开展跨界合作）和可持续发展目标 17（重振可持续发展伙伴关系）等实施各级水资源综合管理的可

194 UNESCO will mobilize and complement the work of various partners including different UN agencies members of UN-Water such as UNEP, UNDP, WMO, UNECE and FAO, etc., academic and research institutions as well as NGOs such as GWP in contributing to promote integrated water resource management. The contribution of IHP-IX will be mainly on capacity development, provision of science-based knowledge, tools, methodologies and guidelines.

PRIORITY 5 : WATER GOVERNANCE BASED ON SCIENCE FOR MITIGATION, ADAPTATION, AND RESILIENCE

195 Water governance refers to the political, social, economic, legal, and administrative systems in place that influence water's access and use, protection from pollution, and management in general. It determines the equity and efficiency in water resource and services allocation and distribution, and balances water use between socio-economic activities and the goods and services provided through ecosystem preservation. It includes formulation, establishment, and implementation of water policies, with clear and practical standards based on science, including water ethics, legislation and institutions, and the roles and responsibilities of all stakeholders. The UNGA resolution A/RES/68/118 on the Law of Transboundary Aquifers, is an example of how UNESCO's scientific support can help in the preparation of water governance related work. The Resolution calls encourages the IHP of UNESCO to continue its contribution by offering further scientific and technical assistance to the States concerned.

196 By 2029, Member States use science-based tools, capacity and knowledge addressing adaptation and mitigation to climate change to significantly reduced water governance gaps.

Relationship between Priority Area 5 and Agenda 2030

197 Good Water Governance is fundamental to the entire concept of setting global goals such as the SDGs and is therefore directly linked to several targets of SDG6 (ensure availability and sustainable management of water and sanitation for all), water-use efficiency (6.4); integrated water resources management (6.5.1), including transboundary cooperation as appropriate (6.5.2); international cooperation and water capacity-building support (6A); and participation of local communities in decision processes (6B). Progress in governance also impacts the fight against poverty (SDG 1) and hunger (SDG 2), building resilience and reducing exposure to extreme weather-related events; doubling productivity and the income of small food producers; and implementing resilient practices that strengthen capacity to adapt to climate change (targets 1.4, 1.5, 2.1, 2.3 and 2.4).

198 There is an important link between SDG 3 (good health and well-being) and water governance, specifically with target 3.9 related to reducing the number of deaths and illnesses from hazardous chemicals, air and water, and contamination. There is also an important connection between this priority area and SDG 4 (quality education), particularly targets 4.1 and 4.5, which aim to eliminate gender disparities and all discrimination in education. Following the same logic, the link with SDG 5 (gender equality) underscores the development of targets to end all forms of discrimination against women and girls and enhance the use of enabling technology to promote the empowerment of women (5.1 and 5.B). It is also linked to SDG 8 (inclusive and sustainable economic growth, employment and decent work for all), specifically with the goal of achieving higher levels of economic productivity through diversification, technological upgrading and innovation and endeavour to decouple economic growth from environmental degradation (targets 8.2, 8.3, 8.4 and 8.9).

199 This priority area strengthens the fulfilment of SDG 10 (reduce inequality within and among countries), SDG 11 (make cities inclusive, safe, resilient and sustainable) and SDG 13 (action to combat climate change and its impacts) and their targets 13.1, 13.2 and 13.B. Sound Governance underpins SDG 16 (peace, justice and strong institutions) and is related to the implementation of integrated water resource management at all levels, including through transboundary cooperation as appropriate (target 6.5) and SDG 17 (partnerships for sustainable development). Any successful sustainable development programme, requires partnerships between governments, the private sector and civil society. Inclusive alliances are built on principles and values, sharing a vision that place people and the planet at the centre of decisions reached. All of this adds up to the fundamental nature of open and good governance as a prime driver in attaining the ambitious targets associated with the SDGs.

持续发展目标相关联。任何成功的可持续发展方案都需要政府、私营部门和民间团体之间的伙伴关系。包容性联盟建立在原则和价值观基础之上，拥有以人类和地球为决策中心的愿景。所有这些都说明了开放和善治的基本特性，是实现与可持续发展目标相关的宏伟子目标的主要动力。

2 预期成果

5.1 支持联合国教科文组织水事机构提高各级决策者对基于科学的水治理的重要性的认识，以便全面提高社区对全球变化影响的韧性。

200 水治理被认为是基石，它能够促进成员国和水领域的利益相关方理解、采纳并实施基于信息和知识的决策，从而建设涵盖所有人的更具韧性和更加繁荣的社区和治理架构。

201 水治理要求人们能够理解和考虑流域及其相关含水层内水资源随降水、蒸散发、下渗和径流等水文循环要素的变化，以及生态系统的主要变化是在哪里以及是如何发生的，以便解决人类住区、农业用水、工业活动等热点问题，并通过干预来避免发生不良变化，促进和改善公平用水或将生态系统恢复到理想状态。

202 由于其特殊性，地下水在有效治理方面仍然落后于地表水，因此仍需做出巨大努力来缩小这一差距。

203 因此，水治理应考虑气候变化等人为因素对生态系统中的水文和营养循环的影响，以科学事实为基础，促进适应性、缓解性和韧性治理。归根结底，充分水治理是保障全人类水安全可持续发展的基础和坚实支柱。

204 政府间水文计划第九阶段战略规划将传播科学成果和最佳实践，并将考虑采取各种定制的和有针对性的方法，包括其他活动（专题讨论会、圆桌会议、研讨会、网络研讨会、讲习班、会议、辩论会、展览）、竞选活动和与媒体接触以提高公众认识的活动，从而基于科学和多学科知识并兼顾全球变化影响，制定关于地表水和地下水的政策手段、工具、决定和行动。

5.2 在水治理工具中融入健全的科学知识，并通过决策者综合考虑地表水和地下水，提高对气候变化和水资源综合管理的适应性。

205 水治理涉及机构角色以及参与水决策的组织和社会团体之间的关系，从横向上看是跨行业和城乡之间的关系，从纵向上看是从地方到国际的关系。水治理需要具有适应性，并通过环境和地点来考虑它的历史和地域的特殊性和挑战性。人们普遍认为治理范畴比政府管辖范围更广，因为治理涉及私营部门、民间团体以及涉及水资源使用和管理的广大利益相关方。

206 城市人口快速增长、特大城市发展、大规模移民和其他因素对实现可持续发展目标 6 构成了挑战，可能危及良好的水治理。其中，影响因素包括资源管理不善、腐败、法律和机构设置不当和运作不良、流域内缺乏合作、官僚主义惰性、人员能力不足、缺乏投资等。

207 联合国教科文组织水事机构将提供科学知识，支持成员国基于科学塑造、改善和更新（如有必要）水治理框架，灵活地缓解和适应水挑战，并恢复抵御全球变化的能力。基于各种相关成果产生的科学知识以及现有的最佳实践，将制定和传播有关方法和指南，促进将可靠的科学更好地融入水治理工具，加强成员国能力。

Expected outputs:

5.1 Awareness raising of decision makers at all levels on the importance of science-based water governance by the UNESCO Water Family supported, to enhance the overall resilience of communities to effects of global change.

200 Water governance is understood as a cornerstone to enable Member States and the multiple stakeholders in water to understand, adopt and implement decisions based on information and knowledge to build more resilient and prosperous communities and governance structures, without leaving anyone behind.

201 Water governance requires the ability to understand and take into consideration what happens to the water resource in a basin and its related aquifer, both in terms of the hydrological cycle (of precipitation, evapotranspiration, infiltration and runoff flows) and where and how the main modifications of ecosystems take place in order to address those hot spots (human settlements, agricultural use, industrial activity, etc.) and to intervene to avoid unwanted modifications, to promote equitable and improved access to water or to rehabilitate ecosystems to a suitable state.

202 Due to its particular characteristics, groundwater still lags behind surface water with respect to effective governance and therefore great efforts are still required to close this gap.

203 Water governance should therefore facilitate adaptation, mitigation and resilience processes considering the human factor, including climate change, impact on the hydrological and nutrient cycles in ecosystems and based on scientific facts. Ultimately, adequate water governance is a fundamental and solid pillar to guarantee sustainable water security for all.

204 IHP-IX will disseminate scientific results, best practices and will consider various tailored and targeted methods including among other events (thematic discussions, roundtables, seminars, webinars, workshops, conferences, debates, exhibitions etc.), campaigns, engaging with media to raise awareness so that policy instruments, tools, decisions and actions on water (surface and underground) are made on the basis of scientific, multidisciplinary knowledge and taking into consideration global change effects.

5.2. Integration of sound science in water governance instruments improved reflecting adaptation to climate change and IWRM, integrating surface and groundwater for their uptake by decision makers.

205 Governance addresses the role of institutions and relationships between organizations and social groups involved in water decision-making, both horizontally across sectors and between urban and rural areas, and vertically from local to international levels. Governance needs to be adaptive, context-dependent, and location-based to take into account historical and territorial specificities and challenges. It is widely accepted that governance is much broader than government as it also seeks to include the private sector, civil society, and the wide range of stakeholders with a stake in water use and management.

206 The fast-increasing rate of urban populations and development of megacities and massive migration and other factors are challenges to the achievement of SDG 6, and thus may jeopardize good water governance. Among them are resource mismanagement, corruption, inappropriate and malfunctioning legal and institutional arrangements, lack of cooperation within river basins, bureaucratic inertia, insufficient human capacity, and a shortage of funding for investments.

207 The UNESCO Water Family will provide scientific knowledge to support Member States to shape, improve and update (if necessary) their water governance frameworks so that they are scientifically based and flexible to respond to the water challenges associated with mitigation, adaptation and resilience to global changes. Building on the scientific knowledge produced within relevant and various outputs and existing best practices, methodologies and guidelines for better integration of reliable science in water governance instruments will be produced and disseminated, and Member States capacity enhanced.

5.3 开展科学评估、制定指导方针，强化“国家自主决定的贡献”和国家适应计划中的涉水内容，以便加强基于水的适应和缓解气候变化政策和行动之间的关系。

208 气候治理是水治理的关键，正如水治理是气候治理的关键一样。政府间气候变化专门委员会从多角度广泛地阐述了理解这一理念的必要性，旨在基于科技和社会层面的不断变化来解决气候变化问题。《2020 年世界水资源发展报告》详述了水和气候的变化，明确提出了水是适应和缓解气候变化方案的一部分。

209 因此，气候缓解和适应政策应更好地结合水问题，并与水政策形成合力。目前，仍然缺乏关于如何将水融入“国家自主决定的贡献”和国家适应计划以及各国如何从更好的水管理中获益的系统性分析。

210 政府间水文计划第九阶段战略规划将通过提供基于科学的分析，支持循证气候政策制定，该分析把目前涉水内容纳入《联合国气候变化框架公约》下的全球气候变化机制，写进“国家自主决定的贡献”和《国家适应计划》章节中，为《联合国气候变化框架公约》关于更好地整合水资源的指导方针提出建议和补充，支持提高定期提交的国家确定贡献和国家适应计划的目标，同时支持《巴黎协定》的衡量、报告和核查机制。通过与《联合国气候变化框架公约》秘书处联络，以及与同样参与制定编写此类补充文件的相关联合国机构和计划合作，确保协调实施这些行动，避免重复工作。

5.4 支持科学界开展和分享关于适应性水资源管理新方法的研究，加强成员国在健全水治理方面的能力。

211 韧性水务部门需要通过新途径采取适应性措施，实现可持续城市水资源管理，而这种管理不能只依赖工程措施和水资源综合管理。它们还需要多方参与、政治意愿以及一个包括战略、战术和业务决策在内的健全科学框架。反过来，这又要求制定战略、行动和监控计划。

212 政策必须确保节约水资源并保护流域，提高对降低耗水的认识，确保遵守法律，管理好含水层补给，回收雨水和废水，特别是在大城市提供循环经济激励措施。这需要国家政府、地方当局、非政府组织、以及其他公、私利益相关方之间的合作。

213 良好的水治理还需要促进更多研究，以便应对缓解和适应气候变化和其他全球变化的挑战；研发低成本技术，为全人类提供解决方案。

214 基于水的多学科方法来考虑解决方案 / 行动对水、环境、生物多样性和人类产生的多重积极影响，应当对已实施的可持续解决方案对流域水资源（环境、土壤、地表和地下水体）和经济方面的影响进行科学的水文分析。

215 政府间水文计划第九阶段战略规划将继续开展和分享关于适应性水管理新方法的研究，特别注意改进关于风险评估、法规、污染控制 / 减轻的科学研究、知识和数据，将水质和水量与经济、社会和生态方法联系起来。政府间水文计划将进一步加强各国专家的能力建设，并组织提高认识的会议，促使联合国教科文组织成员国的决策者们了解这项研究的成果。

5.5 加强科学界和决策者在开发新管理框架和工具方面的能力建设，以便巩固水治理、增强韧性。

216 联合国水机制可持续发展目标 6 综合报告指出，良治对于多个利益相关方参与的自下而上的可持续水

5.3. Science-based assessment and development of guidelines, for strengthening water-related content in Nationally Determined Contributions and National Adaptation Plans, conducted to strengthen water-based climate policy-action nexus for adaptation and mitigation.

208 Climate Governance is essential for Water Governance just as water governance is essential for climate governance. The Intergovernmental Panel on Climate Change (IPCC) has reflected on the need to understand this concept from a broader and different perspective that allows addressing solutions to climate change based on the constant changes that are carried out at a scientific technological and social level. The World Water Development Report 2020 on water and climate change has clearly shown that water is part of the solution to climate change both for adaptation and also for mitigation.

209 Therefore, climate mitigation and adaptation policies should better consider water and make synergy with water policy. A systematic analysis of how water is included in Nationally Determined Contributions and National Adaptation Plans and how countries can benefit from a better water management is still missing.

210 IHP-IX will support evidence-based climate policy making by providing a science-based analysis of current inclusion of water-related content into the global climate change regime under the UNFCCC, including within Nationally Determined Contributions (NDCs) and National Adaptation Plan (NAPs), developing also recommendations and supplements to the UNFCCC work on guidelines for better integrating water and supporting the increased ambition in the periodical re-submission of NDCs and new sub-missions of NAPs, at the same time supporting the Measuring, Reporting and Verification mechanism of the Paris Agreement. These activities will be realized in liaison with UNFCCC Secretariat and in collaboration with relevant UN bodies (agencies and programmes) who are also involved in the development of such supplements, to ensure appropriate harmonization of actions and to avoid duplication of efforts.

5.4. Conducting and sharing of research on novel approaches of adaptive water management by the scientific community supported and capacities of Member States strengthened to enhance sound water governance.

211 Adaptation measures for a resilient water sector require new paths to achieve sustainable urban water management that goes beyond physical engineering and implementing IWRM. They further require the participation of multiple actors, political will and a sound scientific framework including strategic, tactical, and operational decisions. These in turn require the development of strategies, action and monitoring plans.

212 Policies must ensure conservation of the resource and protection of watersheds, raise awareness for the reduction of water consumption, ensure compliance with the law, manage aquifer recharge, and recycle storm water and wastewater, provide circular economy incentives, especially in megacities. This requires cooperation of the national governments, local authorities and non-governmental organizations, as well as other public and private stakeholders.

213 Good water governance also requires the promotion of additional research to address the challenges of mitigation and adaptation to climate and other global changes as well as to develop affordable technologies for providing solutions to all.

214 Scientific hydro analyses on the impact of implemented sustainable solutions in terms of water resources (environment, soil, surface and groundwater bodies) and economics in watersheds, considering a multi-disciplinary approach based on water positive multi-benefit impact of solutions/actions, for water, environment, biodiversity and people, should be performed.

215 IHP-IX will continue conducting and sharing of research on novel approaches of adaptive water management, paying special attention to improving scientific research, knowledge and data on risk assessment, regulations, pollution control/attenuation, linking water quality and quantity with economic, societal and ecological approaches. IHP will further build capacity of national experts and organize awareness raising sessions to sensitize decision makers within UNESCO's Member States to the results of this research.

5.5 Capacities of the scientific community and decision makers strengthened on new frameworks and tools, to underpin water governance and build resilience.

216 The UN-Water SDG 6 Synthesis Report suggests that good governance is essential for sustainable water management focusing on a bottom-up framework with multiple stakeholders. It's necessary to improve decision-making processes through public participation that "can ensure that decisions are based on shared knowledge, experiences and scientific evidence, are influenced by the views and experience of those affected by them, that innovative and creative options are considered and that new arrangements are workable, and

资源管理至关重要。为了改善决策过程，有必要通过公众参与“确保基于共同知识、经验和科学证据进行决策，并在决策过程中考虑受影响者的意见和经验，考虑创新性和创造性方案，同时确保新方案的可行性和公众认可度”（欧洲环境署，2015 年第 12 号技术报告）。

217 既然决定应对水挑战，那么需要进一步采纳全面的、一致的和跨部门的观点，采用基于科学的政策来解决水资源不确定性所涉及的方方面面。为了构建抵御不确定性和未来风险的能力，需要成员国所有利益相关方在一个有利的法律环境、科学和体制框架内持续合作。

218 基于多学科（水文、社会和社会经济科学等）的科学方法将为各级水治理提供一个新的全面性框架。向流域内城市和农村地区所有利益相关方分享科学知识和对水问题的理解，将有利于制定共同愿景、实施基于科学的解决方案。考虑到全球变化带来的挑战，政府间水文计划第九阶段战略规划将采用参与式方法，如：气候风险知情决策分析、水敏感城市设计等，继续开发新框架和工具、巩固现有框架和工具，支持水治理并激发科学界、决策者的活力。还将进一步举办培训班和提高认识的会议，加强科学家和决策者吸收和实施这些创新方法、框架和工具的能力。

3 与其他联合国机构和科学伙伴的合作

219 联合国教科文组织在水治理方面的作用是加强由科学知识架构的决策和政策制定的科学基础，考虑全球变化的影响。

220 通过与科学联盟、政府间组织、非政府组织和国际金融机构等其他联合国机构和伙伴合作，政府间水文计划第九阶段战略规划将促进基于科学的水治理的改善，建设韧性社会。

221 针对含跨界水资源在内的所有水资源，政府间水文计划将与以下现有倡议和组织合作，提供科学知识和研究技能。这些倡议和组织有：联合国开发计划署、联合国欧洲经济委员会、联合国环境规划署、联合国气候变化框架公约、国际原子能机构、世界气象组织、经济合作与发展组织、斯德哥尔摩国际水资源研究所以及全球水伙伴等。

acceptable to the public" (EEA, 2015, 12).

217 Decisions to deal with water challenges require further to a holistic, coherent, and inter-sectoral vision, science-based policies to address all aspects of water uncertainty. Building resilience to uncertainty and future risks requires a continuous partnership of all stakeholders in Member States, working within an enabling legal, scientific, and institutional framework.

218 A multidisciplinary science approach (hydrological, social and socio-economics sciences, etc.) will provide a new holistic framework for water governance at all levels. Sharing scientific knowledge and understanding of water issues to all stakeholders, in urban areas and in rural areas in a watershed will allow the development of a common vision and implementation of science-based solutions. IHP-IX will continue to develop new and strengthen existing frameworks and tools with a participatory approach, such as Climate Risk Informed Decision Analysis (CRIDA), Water Sensitive Urban Design, etc., taking into consideration the challenges posed by global changes, to underpin water governance and build resilience of the scientific community and decision makers. It will further organize training and raising awareness sessions to strengthen the capacities of scientists and decision makers to enhance the uptake and implementation of these innovative approaches, frameworks and tools.

Cooperating with other UN Agencies and scientific partners

219 UNESCO's role in water governance is to reinforce the scientific base upon which decisions and policies are framed by providing scientific knowledge, which will take into consideration the effects of global change.

220 IHP-IX will cooperate with other UN agencies and other partners including Scientific Unions, Intergovernmental Organizations, Non-Governmental Organizations and International Financial Institutions and others in contributing to improve water governance based on science for building resilient societies.

221 IHP will provide scientific knowledge and research skills cooperating with existing initiatives and Organizations, such as UNDP, UNECE, UNEP, UNFCCC, IAEA, WMO, OECD, SIWI as well as GWP and others, in the case of all water resources, including transboundary ones.

第四章
政府间水文计划
第九阶段战略规划
实施手段和实施路径

IMPLEMENTATION MEASURES AND WAY FORWARD OF IHP-IX

第一节　沟通和外联

222 有效沟通和外联是政府间水文计划第九阶段战略规划不可或缺的关键部分。政府间水文计划第九阶段战略规划实施者之间的沟通流程是有效实现该计划的先决条件。联合国教科文组织水事机构成员及伙伴需要通过可靠的网络和沟通能力开展积极合作，这是落实政府间水文计划第九阶段战略规划的关键工具。政府间水文计划水信息网络系统为此类合作提供了关键的有利条件。

223 除了沟通和外联工作外，政府间水文计划第九阶段战略规划还致力于加强政府间水文计划水事机构网络、其他利益相关方及伙伴的协作参与，提高第九阶段战略规划的公众知名度和认可度，及其在促进全球水安全方面的作用。

224 政府间水文计划第九阶段战略规划系统地整合了各级实施过程中的沟通和外联工作，包括反馈机制，以便利用该计划的产出和成果来加强并展示其影响。有效实施政府间水文计划的外联和沟通战略是提高政府间水文计划第九阶段战略规划知名度的关键，有助于动员更多的伙伴和资金，从而有效执行计划。

第二节　实施路径

225 政府间水文计划的第九阶段将涵盖直至《2030 年可持续发展议程》结束的未来八年。其目标是支持成员国实现可持续发展目标 6、其他涉水可持续发展目标和国际涉水议程，建立一个水安全世界和韧性社会。实施该计划将有助于联合国可持续发展目标 6“全球加速框架”和联合国《水行动十年宣言》（2018—2028 年）的实施。应该基于成员国在解决全球复杂的相互联系和水挑战方面所取得的进展来衡量该计划的成功与否，即人民和机构拥有足够的能力和科学基础知识，通过对水管理和水治理作出明智决策，实现可持续发展并建设韧性社会。

226 政府间水文计划第九阶段战略规划的全面实施将由联合国教科文组织水事机构领导，并与联合国水机制及其成员、学术和科学组织和协会、政府间组织、区域或国家组织、非政府组织、全球基金、研究 / 学术界和私营部门等伙伴开展合作。

227 经政府间水文计划理事会批准本战略规划后，将制定指导实施本战略规划的业务实施计划。业务实施计划将用于指导逐步实现本战略规划，并根据实施过程中的主要经验教训和不断变化的环境、能力与可用资源进行调整。业务实施计划将确定每个优先领域的具体活动和交付成果，包括每个里程碑式的成果。

228 业务计划还将确定各项产出 / 交付成果的责任主体。将为每项特定的产出 / 成果制定具体的衡量指标和核查手段，使不同利益相关方能够获取相关信息和数据，以便进行明确的监控和评估。将定期审查业务计划，并将审查报告呈给执行局和政府间水文计划理事会会议，评估成绩和挑战，吸取教训并调整剩余业务计划。

OUTREACH AND COMMUNICATION

222 Effective communication and outreach are integral key components of IHP-IX. The communication flow among the implementing actors of IHP-IX is a prerequisite for the Programme's impactful realization. The proactive cooperation of the members of the UNESCO Water Family and its partners requires reliable networking and communication capacities as key tools for IHP-IX. IHP-WINS provides a pivotal asset in this regard.

223 Communication and outreach of IHP-IX endeavour, simultaneously, the strengthening of the collaborative engagement of IHP's family network and other stakeholders and partners and increased public visibility and recognition of IHP-IX and its role in contributing to global water security.

224 IHP-IX systematically integrates communication and outreach in its implementation at all levels, including feedback mechanisms which will allow harnessing the Programme's outputs and outcomes both to enhance and demonstrate impact. Effective implementation of the IHP outreach and communication strategy is crucial in increasing the visibility of IHP-IX and contributing to mobilize more partnerships and funding for an impactful implementation of the programme.

WAY FORWARD

225 This ninth phase of IHP-IX will cover the next eight years almost until the end of agenda 2030. It is designed to support Member States in achieving SDG6 and other water-related SDGs and international water-related agendas towards a water-secure world and resilient societies. Its implementation will among others contribute to the UN SDG6 Global Accelerator Framework and the UN Water Action Decade (2018—2028). Its success should be measured based on the progress made by Members States in addressing global complex interlinkages and water challenges by people and institutions having adequate capacity and scientifically based knowledge to make informed decisions on water management and governance to attain sustainable development and to build resilient societies.

226 The overall implementation of IHP-IX will be led by UNESCO's Water Family and will benefit from partnerships with UN-Water and its members, academic and scientific organizations and associations, intergovernmental organizations, regional, or national organizations, Non-Governmental organizations, global funds, research/academia and the private sector.

227 The implementation of this strategic plan will be guided by an Operational-Implementation Plan, to be elaborated after the approval of the strategic plan by the IHP council. It will guide the incremental achievements of the strategic plan, grounded in key lessons learned during implementation and adjustments to changing contexts, capacities and resource availability. Specific activities and deliverables including milestones in line with each of the outputs under each priority areas will be identified.

228 The operational plan will also identify the entities responsible for the respective outputs/deliverables. Specific measurable indicators will be developed for each of the specified outputs/deliverables and the means of verification to enable the different stakeholders capture the relevant information and data for a clear monitoring and evaluation. Regular reviews of the operational plan will be performed to be presented at the bureau and IHP council meetings to take stock of the achievements and challenges, and to draw lessons allowing adjustments in the remaining part of the plan.

229 将动员所有成员国以及政府间水文计划国家委员会、二类中心、教席、旗舰计划 / 倡议、世界水资源评估计划以及主要合作伙伴等联合国教科文组织水事机构共同编制业务实施计划。根据政府间水文计划的政府间性质和第八阶段战略规划中期外部评估建议，成员国及其相关的联合国教科文组织水事机构实体应在实施业务计划中发挥关键作用，以便确定可对哪些产出作出积极贡献和 / 或对哪些产出发挥（共同）领导作用。根据政府间水文计划新章程，秘书处将为此提出明确的实施框架，包括设立专题实施工作组，供政府间水文计划理事会讨论和批准。还将制定融资战略，以确保政府间水文计划第九阶段战略规划的实施。

229 All Members States and the UNESCO Water Family, comprising the national IHP committees, the centres, chairs, flagships/initiatives and WWAP and key partners will be mobilized in the preparation of the Operational-Implementation Plan. In line with the intergovernmental nature of IHP and the recommendations of the mid-term external evaluation of IHP-VIII, Member States and their related UNESCO Water Family entities should play a key role in the implementation of the operational plan by identifying outputs for which they can actively contribute to and/or provide (co-)leadership. A clear implementation framework will be proposed by the Secretariat in that regard in line with the new IHP statutes including among others the establishment of thematic implementation working groups to be discussed and approved by the IHP council. A Financing Strategy will also be prepared to ensure implementation of IHP-IX.

附件

ANNEXES

1 术语表

术 语	释 义
公民科学	公民科学指公民在专业科学家经常性指导下和/或与科学机构或正式科学计划合作，通过遵循科学有效的方法，自愿参与科学研究和数据收集。 参照并根据《联合国教科文组织开放科学建议书》第 10 段 [有待列入确切参考资料]。
生态水文学	生态水文学是水文学的一个分支，强调不同尺度水文、生物和生态过程之间的关系，并将其转化为基于自然的解决方案来提高水安全；通过协调社会经济需求和环境潜力，增强生物多样性并创造可持续发展机会。 [https://en.unesco.org/themes/water-security/hydrology/ecohydrology]
全面水资源管理	全面水资源管理指水管理的实践和解决方案，人类对整个水系统、循环及其资源进行处理或治理，而不仅仅处理或治理其中一部分。
水文信息学	水文信息学是一个基于数学模型的研究领域，通过集成信息和通信技术进行数据采集、建模和决策支持，考虑对水生环境和社会的影响以及对基于水系统的管理，研究现实世界中水动态信息流并凝练相关知识。 [联合国教科文组织国际水利环境工程学院，即目前的荷兰代尔夫特国际水利环境工程学院，www.unesco-ihe.org]
水资源（水）政策	水资源（水）政策包括政策制定过程，该过程将影响水的收集、处理、使用和处置，以便支持人类用水、保护环境质量。水政策涉及供水、用水、水处理和可持续发展决策。 [https://www.wikiwand.com/en/Water_resource_policy]
水利工程	水利工程是科学技术的一个分支，研究水资源及其各种用途，预防水的破坏性影响。 [https://www.epictraining.ca/course-catalogue/civil/10453/hydrotechnical-engineering-for-non-hydrotechnical-engineers]
包容性水资源管理	包容性水资源管理是一种综合水（资源）管理实践，旨在包括和集合不同人群，并让他们参与水资源管理活动、组织、政治进程，尤其关注那些处于劣势、遭受歧视人群或残障人群。
水资源综合管理	世界气象组织和联合国教科文组织将其定义为：水资源综合管理是指区域水资源的开发和运营，兼顾水文、技术、社会经济、政治和环境多重要素。 全球水伙伴将其定义为：水资源综合管理是促进水、土地和相关资源的协调发展和管理的过程，旨在公平地、最大限度地提高经济和社会福利，且不影响重要生态系统的可持续性。
基于自然的解决方案	基于自然的解决方案指保护、可持续管理和恢复天然或改良的生态系统的行动，旨在有效应对气候变化、粮食安全、灾害风险、水安全、社会经济发展以及人类健康等共同的社会挑战，同时造福于人类福祉和生物多样性。 [Cohen-Shacham，E.，G. Walters，C. Janzen，S. Maginnis （eds）. 2016. 全球社会挑战应对之道：基于自然的解决方案。 瑞士格兰：国际自然保护联盟。下载网址： https://portals.iucn.org/library/node/46191]
非常规水资源	非常规水资源指海水淡化等专门工艺附带产生的水资源；或在使用前需要经过适当处理的水资源；或用于灌溉时需要相关农场管理的水资源；或需要使用特殊技术来收集 / 获取的水资源。 [https://inweh.unu.edu/projects/unconventional-water- resources/]
开放型科学 开放型数据	开放型科学背后的理念是在所有利益相关方积极参与的情况（向社会开放）下，使社会公众能够更广泛地获取（开放型存取）并更可靠地利用科学信息、数据和产出（开放型数据）。 [https://en.unesco.org/science-sustainable-future/open-science]

GLOSSARY

Term	Definition
Citizen Science	With reference to and in line with the UNESCO Open Science Recommendation, para. 10 [exact reference to be included]: Citizen Science is the voluntary participation of citizens in scientific research and data collection, frequently under the direction of professional scientists and/or in association with scientific institutions or formal scientific programmes, following scientifically valid methodologies.
Ecohydrology	A branch of hydrology, which highlights the relationships between hydrological, biological and ecological processes at different scales and translates them into nature–based solutions to improve water security, enhance biodiversity and create opportunities for sustainable development by harmonizing socio–economic needs and environmental potential. https://en.unesco.org/themes/water–security/hydrology/ecohydrology
Holistic Water Management	Water management practices and solutions where one deals with or treats the whole of the water system, cycle and its resources, not just a part.
Hydro–informatics	A mathematical model–based field of study of the flow of information and the generation of knowledge related to the dynamics of water in the real world, through the integration of information and communication technologies for data acquisition, modeling and decision support, and taking into account the consequences for the aquatic environment and society and for the management of water–based systems.(UNESCO–IHE, currently IHE–Delft, www.unesco–ihe.org)
Hydro (Water) Policy	Water resource policy encompasses the policy–making processes that affect the collection, preparation, use and disposal of water to support human uses and protect environmental quality. Water policy addresses provision, use, disposal and sustainability decisions. https://www.wikiwand.com/en/Water_ resource_policy
Hydrotechnical Engineering	A branch of science and technology concerned with the study of water resources and their use for various purposes, as well as the prevention of the damaging effect of water. https://www.epictraining.ca/course– catalogue/civil/10453/hydrotechnical–engineering–for–non–hydrotechnical–engineers
Inclusive Water Management	Integrated water (resources) management practices aiming to include, integrate, and have the participation of all diverse people and groups in its activities, organizations, political processes, etc., with special attention to those who are disadvantaged, have suffered discrimination, or are living with disabilities.
Integrated Water Resources Management	Development and operation of regional water resources, taking into account hydrological and technical aspects, as well as socio–economic, political and environmental dimensions. (WMO UNESCO Glossary) A process which promotes the coordinated development and management of water, land and related resources in order to maximize the resultant economic and social welfare in an equitable manner without compromising the sustainability of vital ecosystems. (Global Water Partnership)
Nature–Based Solutions (NBS)	Actions to protect, sustainably manage, and restore natural or modified ecosystems, that address societal challenges effectively and adaptively, simultaneously providing human well–being and biodiversity benefits, with climate change, food security, disaster risks, water security, social and economic development as well as human health being the common societal challenges. (Cohen–Shacham, E., G. Walters, C. Janzen, S. Maginnis (eds). 2016. Nature–based solutions to address global societal challenges. Gland, Switzerland: IUCN. Xiii + 97 pp. Downloadable from https://portals. iucn.org/library/node/46191)
Non–conventional Water Resources	Water resources that are generated as a by–product of specialized processes such as desalination; or that need suitable pre–use treatment; or pertinent on–farm management when used for irrigation; or need a special technology to collect/access water. (https://inweh.unu.edu/projects/unconventional–water–resources/)
Open Science Open Data	The idea behind Open Science is to allow scientific information, data and outputs to be more widely accessible (Open Access) and more reliably harnessed (Open Data) with the active engagement of all the stakeholders (Open to Society). (https://en.unesco.org/science–sustainable–future/open–science)

社会水文学	社会水文学是一个研究水–人动态交互和反馈的跨学科领域，涉及领域包括对水文和社会过程的相互作用的历史研究，不同文化中人类和水系统的共同演化和自我组织的比较分析，以及基于过程的人–水系统的耦合模型。 [Sivapalan, M., Savenije, H. H. G, Blöschl, G.（2012-04-15）. 社会水文学：一门关于人类和水的新科学：特邀评论 .《水文过程》，26（8）：第 1270–1276 页 . doi：10.1002/hyp.8426]
可持续水资源管理	可持续水资源管理指在不损害后代开发水资源能力的前提下，满足当代用水需求的能力。 [https://www.aquatechtrade.com/news/water-treatment/sustainable-water-essential-guide/#what-is-water-sustainability]
跨学科 / 跨学科研究	跨学科研究是通过整合多个科学学科和学术界以外的各种利益相关方的知识，生成有关复杂社会问题知识的一种研究模式。针对共同定义的问题开展跨学科研究，采取具体情境方法和自我反思方法，生成对社会和科学都有价值的三类知识：（1）关于人类–环境系统的系统知识；（2）关于人类–环境系统预想状态的目标知识；（3）关于如何达到预想状态的转化知识。 [Jahn，T. Bergmann，M.，Keil，F.（2012）：跨学科：主流化与边缘化之间。《生态经济》，第 79 卷，1-10.]
城市代谢	城市代谢被定义为“城市出现的技术和社会经济过程的总和，导致城市增长、能源生产和废物清除……（考虑）水、材料、能源和营养物等四类基本流或循环”。（Kennedy, Cuddihy et al, 2007） 循环型城市代谢是指将循环经济原则与城市代谢方法相结合。
水文化	剑桥词典将其定义为：水文化是指特定时间内特定群体在生活中与水的关系，尤其指一般的习俗和信仰。 [https://dictionary.cambridge.org/dictionary/english/culture] Your Dictionary 词典将其定义为：水文化是与特定地理区域、文化群体、宗教或遗产要求等有关的特定水资源的价值和需求。 [https://www.yourdictionary.com/cultural-water] * 植物在水中生长，（后来）特指植物在具有或（通常）没有固体基质的人工系统中生长。 [https://www.lexico.com/definition/water_culture]
水治理	全球水伙伴将其定义为：水治理是为开发和管理水资源并在社会不同阶层提供水服务而设立的一系列正式和非正式的政治、社会、经济和行政制度。 [GWP. GLOBAL WATER PARTNERSHIP. www.gwpforum.org/servlet/PSP] 经济合作与发展组织将其定义为：水治理是一套规则、实践以及（正式与非正式）程序，并通过其制定和实施水资源和服务管理的决策，利益相关方会表达他们的需求，并对决策者进行问责。 （经济合作与发展组织，2015 年）
水资源管理总体规划	水资源管理总体规划是为了有序且综合地规划和实施水资源计划和项目，并根据国家社会经济发展总体目标而建立合理管理水资源的基本框架。 [《国家水资源总体规划编制指南》《水资源丛书》第 65 期，联合国，纽约，1989 年]
水资源紧张和缺水	每年人均可用的可再生水资源数量是根据对家庭、农业、工业和能源部门的用水需求以及环境需水量来估算的。当一个国家的可再生水资源供应量不能维持在 1 700 立方米时，这个国家就会出现水资源紧张的情况；当供应量低于 1 000 立方米时，这个国家就会出现缺水现象；当供应量低于 500 立方米时，这个国家则会出现严重缺水现象。(Falkenmark，Lundqvist and Widstrand，1989) [联合国教科文组织 2011 年：亚洲及太平洋地区道德与气候变化项目，第 14 工作组报告：《水资源道德与水资源管理》，2011 年；https://unesdoc.unesco.org/ark:/48223/pf0000192256] [Rijsberman，F. R. 2005. 缺水：事实还是虚幻?《第四届国际作物科学大会论文集》，2004 年 9 月 26 日至 10 月 1 日，澳大利亚布里斯班。以光盘形式出版。网站：www.cropscience.org]
水利益相关方	水利益相关方是对涉水项目、计划或政策举措的特定部门或体制或结果有一定需求的某一机构、组织或团体。 [欧洲环境署网站：www.eea.eu.int；词汇表见：eea.eu.int/EEAGlossary/]

Socio-hydrology	An interdisciplinary field studying the dynamic interactions and feedbacks between water and people, in areas such as the historical study of the interplay between hydrological and social processes, comparative analysis of the co-evolution and self-organization of human and water systems in different cultures, and process-based modelling of coupled human-water systems. (Sivapalan, Murugesu; Savenije, Hubert H. G.; Blöschl, Günter (2012-04-15). "Socio-hydrology: A new science of people and water: INVITED COMMENTARY" . Hydrological Processes. 26 (8): 1270 - 1276. doi:10.1002/hyp.8426.)
Sustainable Water Management	Sustainable water management means the ability to meet the water needs of the present without compromising the ability of future generations to do the same. (https://www.aquatechtrade.com/news/water-treatment/sustainable-water-essential-guide/#what-is- water-sustainability)
Transdisciplinarity/ Transdisciplinary Research	A research mode for generating knowledge about complex societal problems by integrating knowledge of multiple scientific disciplines and various stakeholders from outside of academia. Transdisciplinary research on a jointly defined problem generates knowledge that is both valuable for society and science. It aims at three types of knowledge, 1) system knowledge about the human-environment system, 2) goal knowledge about the envision state of the system and 3) transformation knowledge about how to reach the envisioned state. Transdisciplinary research requires context-specific methods and a self-reflexive approach. (Jahn, T. Bergmann, M., Keil, F. (2012): Transdisciplinarity: Between mainstreaming and marginalization. Ecological Economics, Vol. 79, 1-10)
Urban Metabolism	Urban Metabolism is defined as "the sum total of the technical and socio-economic process that occur in cities, resulting in growth, production of energy and elimination of waste··· [taking into account] four fundamental flows or cycles, those of water, materials, energy, and nutrients" . (Kennedy, Cuddihy et al. 2007) Circular urban metabolism refers to the combination of circular economy principles to the urban metabolism approach.
Water Culture	The way of relations to water in life, especially the general customs and beliefs, of a particular group of people at a particular time (https://dictionary.cambridge.org/dictionary/english/culture) The definition of cultural water means the values and demands of specific water sources that relate to a specific geographic area, a cultural group, a religion or heritage requirement, etc. (https://www.yourdictionary.com/cultural-water) *The growing of plants in water, (in later use) especially in an artificial system with or (usually) without a solid substrate. (https://www.lexico.com/definition/water_culture)
Water Governance	The range of formal and informal political, social, ecnomic and administrative systems that are in place to develop and manage water resources, and the delivery of water services, at different levels of society (GWP, GLOBAL WATER PARTNERSHIP. www.gwpforum.org/servlet/PSP) Water governance is the set of rules, practices, and processes (formal and informal) through which decisions for the management of water resources and services are taken and implemented, stakeholders articulate their interest, and decision-makers are held accountable (OECD, 2015a).
Water Management Master Plan	Water Master Plan is to establish a basic framework for: orderly and integrated planning and implementation of water resources programmes and projects; and a rational water resources management consistent with overall national socioeconomic development objectives. (Guidelines for the preparation of National Water Master Plans, Water Resources Series No. 65, UN, N.Y. 1989)
Water Stress & Water Scarcity	The amount of available renewable water resources per capita per year based on estimates of water requirements in the household, agricultural, industrial and energy sectors, and the needs of the environment. Countries whose renewable water supplies cannot sustain 1 700 m^3 are said to experience water stress. When supply falls below 1,000 m^3 a country experiences water scarcity, and below 500 m^3 absolute scarcity. Falkenmark, Lundqvist and Widstrand, 1989 UNESCO 2011: Ethics and Climate Change in Asia and the Pacific (ECCAP) Project, Working Group 14 Report, Water Ethics and Water Resource Management, 2011; https://unesdoc.unesco.org/ark:/48223/pf0000192256 Rijsberman, F. R. 2005. Water Scarcity: Fact or Fiction? Proceeding of the 4th International Crop Science Congress, 26 September-1 Octobor 2004. Brisbane, Australia. Published in CD-Rom. Website: www.cropscience.org)
Water Stakeholder	An institution, organization or group that has some interest in a particular sector or system or outcome of a project, programme or policy initiative related to water (EEA. European Environmental Agency. www. eea.eu.int/; glossary: eea.eu.int/EEAGlossary/)

2 伙伴清单（暂定）

1） 联合国实体、政府间组织

联合国粮食及农业组织、全球环境基金、联合国经济和社会事务部、联合国开发计划署、联合国开发计划署－全球环境基金、联合国减少灾害风险办公室、联合国欧洲经济委员会、联合国环境规划署、联合国环境规划署－全球环境基金、联合国人居署、联合国儿童基金会、联合国外层空间事务厅、联合国大学、世界卫生组织、世界气象组织、联合国非洲经济委员会、联合国欧洲经济委员会、联合国亚洲及太平洋经济社会委员会、联合国拉丁美洲和加勒比经济委员会、联合国西亚经济社会委员会。

非洲水资源部长理事会、经济合作与发展组织。

2） 国家和国际组织

国家和国际流域组织。

3） 专业科学组织

国际船级社协会、国际水文地质学家协会、国际水利与环境工程学会、国际水文科学协会、创新用户群、国际水资源协会、世界冰川监测服务处。

4） 学术机构、研究中心和计划

36 个涉水二类中心、66 个联合国教科文组织涉水教席。

农业气象和水文气象训练方案、全球水未来、家庭用水不安全体验、斯德哥尔摩国际水研究所、全球能源与水资源交换计划、国际水资源管理研究所。

5） 公约

《国际水道非航行使用法公约》（纽约，1997 年）、《跨界水道和国际湖泊保护与利用公约》（赫尔辛基，1992 年）、《联合国气候变化框架公约》、《联合国防治荒漠化公约》、《国际重要湿地公约》（拉姆萨尔，1971 年）。

6） 非政府组织

世界水理事会、全球水适应联盟、未来之水、全球水伙伴、全球水伙伴－地中海委员会、水安全管理评估研究和技术协会。

7） 私营部门

盖洛普咨询有限公司、各供水公司。

Indicative List of Partners (provisional)

UN entities, Intergovernmental Organizations:

FAO; GEF; UNDESA; UNDP, UNDP-GEF; UNDRR; UNECE; UNEP, UNEP-GEF; UN-HABITAT; UNICEF; UNOOSA; UNU; WHO; WMO; UN Economic Commission for Africa (UNECA), UN Economic Commission for Europe (UNECE), UN Economic and Social Commission for Asia and the Pacific (UNESCAP), UN Economic Commission for Latin America and the Caribbean (UNECLAC), UN Economic and Social Commission for Western Asia (UNESCWA)

The African Ministers' Council on Water (AMCOW); OECD

National and international organizations:

National and international river basin organizations

Professional scientific organizations:

IACS; IAH; IAHR; IAHS; IUGG; IWRA; WGMS

Academic institutions, research centres and programmes:

All 36 Category 2 water related Centres and 66 UNESCO Water related Chairs

AGRHYMET; Global Water Future (GWF); HWISE; SIWI, Global Energy and Water Exchanges (GEWEX); IWMI

Conventions:

Convention on the Law of the Non-navigational Uses of International Watercourses (New York, 1997); Convention on the Protection and Use of Transboundary Watercourses and International Lakes (Helsinki, 1992); UNFCCC; UNCCD; Convention on Wetlands of International Importance (Ramsar, 1971)

Non-governmental Organizations:

World Water Council (WWC); Alliance for Global Water Adaptation (AGWA); Future Water; GWP, GWP-Med; W-SMART

Private sector:

Gallup; various water utilities

3　现代水文学的主要问题 *

1）时间变异和未来变化

（1）气候和环境变化条件下的区域水文循环是否正在加速 / 减速？是否存在区域水文循环加速 / 减速的临界点（或不可逆变化）？

（2）寒冷地区径流和地下水如何随气候变暖而变化（如冰川融化和永久冻土融化）？

（3）气候变化和用水改变（半）干旱地区间歇性河流和地下水的机制是什么？

（4）土地覆盖变化和土壤扰动对地表水和能量通量以及由此产生的地下水补给有什么影响？

2）空间变异和尺度问题

（5）造成径流、蒸发、地下水以及物质通量（碳与其他营养物质，沉积物）的空间异质性和同质性的原因是什么？以及这些物质通量对控制因素（例如降雪规律、干旱度、反应系数）有怎样的敏感性？

（6）流域尺度上的水文规律是什么？这些规律如何随着尺度变化？

（7）为什么大多数优先流能跨越多个尺度？这种行为如何与关键带共同演化？

（8）为什么当暴雨径流是经年之水时河流对降水输入的响应如此之快？陆地水循环中水的过境时间分布是多少？

3）极端事件的变异

（9）洪水多发期和干旱多发期是怎样形成的？它们是否发生变化？如果是，那么为什么会发生这样的变化？

（10）为什么有些流域的径流极值对土地利用 / 覆盖和地貌变化比其他流域更敏感？

（11）为什么积雪降雨事件会产生极端径流？这种特大径流的发生机制是什么？什么时间发生？

4）水文界面

（12）控制山坡—河畔—河流—地下水相互作用的过程是什么？这些过程在何时相互连接？

（13）控制跨界地下水通量（例如地下水补给量、流域间通量和入海流量等）的过程是什么？

（14）哪些因素导致了水质的长期恶化？

（15）新兴污染物的范围、宿命和影响是什么？如何清除或灭活地下微生物病原体？

5）测量和数据

（16）如何利用创新技术测量不同时空尺度的地表和地下水的属性、状态和通量？

（17）与软数据（非专业人员和数据挖掘等定性观测）相比，传统水文观测的相对价值是什么？在什么条件下可用海量空间数据代替长期观测数据？

（18）我们如何从人类和水系统的现有数据中提取信息，以便辅助社会 - 水文概念和模型构建过程？

6）建模方法

（19）如何调整水文模型，使其能够推断不断变化的条件，包括植被的动态变化？

---

* 根据 Blöschl，G. 等人（2019 年）《水文学的二十三个未解决问题——水文界观点》，《水文科学杂志》，第 64 卷第 10 期，第 1141–1158 页。

ANNEX 3: MAIN QUESTIONS OF MODERN HYDROLOGY*

Time variability and change

1. Is the hydrological cycle regionally accelerating/decelerating under climate and environmental change and are there tipping points (irreversible changes)?
2. How will cold region runoff and groundwater change in a warmer climate (e.g. with glacier melt and permafrost thaw)?
3. What are the mechanisms by which climate change and water use alter ephemeral rivers and groundwater in (semi) arid regions?
4. What are the impacts of land cover change and soil disturbances on water and energy fluxes at the land surface, and on the resulting groundwater recharge?

Space variability and scaling

5. What causes spatial heterogeneity and homogeneity in runoff, evaporation, subsurface water and material fluxes (carbon and other nutrients, sediments), and in their sensitivity to their controls (e.g. snow fall regime, aridity, reaction coefficients)?
6. What are the hydrologic laws at the catchment scale and how do they change with scale?
7. Why is most flow preferential across multiple scales and how does such behaviour co-evolve with the critical zone?
8. Why do streams respond so quickly to precipitation inputs when storm flow is so old, and what is the transit time distribution of water in the terrestrial water cycle?

Variability of extremes

9. How do flood-rich and drought-rich periods arise, are they changing, and if so why?
10. Why are runoff extremes in some catchments more sensitive to land-use/cover and geomorphic change than in others?
11. Why, how and when do rain-on-snow events produce exceptional runoff?

Interfaces in hydrology

12. What are the processes that control hillslope-riparian-stream-groundwater interactions and when do the compartments connect?
13. What are the processes controlling the fluxes of groundwater across boundaries (e.g. groundwater recharge, inter-catchment fluxes and discharge to oceans)?
14. What factors contribute to the long-term persistence of sources responsible for the degradation of water-quality?
15. What are the extent, fate and impact of contaminants of emerging concern and how are microbial pathogens removed or inactivated in the subsurface?

Measurements and data

16. How can we use innovative technologies to measure surface and subsurface properties, states and fluxes, at a range of spatial and temporal scales?
17. What is the relative value of traditional hydrological observations vs soft data (qualitative observations from lay- persons, from data mining etc.), and under what conditions can we substitute space for time?
18. How can we extract information from available data on human and water systems in order to inform the building process of socio-hydrological conceptualisations and models?

Modelling methods

19. How can hydrological models be adapted to be able to extrapolate to changing conditions, including changing vegetation dynamics?

---

*(according to Blöschl, G. et al. (2019) Twenty-three unsolved problems in hydrology (UPH) – a community perspective, Hydrological Sciences Journal, 64:10, 1141–1158)

（20）我们如何厘清并降低水文预测中模型结构 / 参数 / 输入的不确定性？

7）与社会接轨

（21）如何将水文预测中的（不）确定性传达给决策者和公众？

（22）与水管理相关的社会目标（例如水—环境—能源—粮食—健康）之间的协同作用和权衡是什么？

（23）水在移民、城市化和人类文明动态中的作用是什么？对当代水管理有什么启示？

20. How can we disentangle and reduce model structural/parameter/input uncertainty in hydrological prediction?

Interfaces with society

21. How can the (un) certainty in hydrological predictions be communicated to decision makers and the general public?
22. What are the synergies and tradeoffs between societal goals related to water management (e.g. water-environment-energy-food-health)?
23. What is the role of water in migration, urbanisation and the dynamics of human civilisations, and what are the implications for contemporary water management?

4 政府间水文计划旗舰倡议清单

1）国际实验和网络数据水流情势（FRIEND-Water）

一个国际研究倡议。旨在通过在区域一级交换数据、知识和技术，帮助设立用于分析水文数据的区域网络。

2）人类活动和气候变化压力下地下水资源评估（GRAPHIC）

一个联合国教科文组织领导的项目。旨在寻求改善对下列机制的理解：（1）全球水文循环内地下水如何相互作用？（2）地下水如何支持人类活动和生态系统？（3）地下水如何响应人类活动和气候变化的复杂双重压力？

3）干旱地区水与发展信息全球网络（G-WADI）

一个关于干旱和半干旱地区水资源管理的全球网络。主要目标是在干旱和半干旱地区建立一个有效促进国际和区域合作的全球性社区。

4）环境、生命和政策水文计划（HELP）

一个综合流域管理新方法。旨在通过构建一个框架，让涉水法律和政策专家、水资源管理者和水科学家共同致力于解决涉水问题。

5）国际干旱倡议（IDI）

一个国际网络平台倡议。旨在为积极开展抗旱工作的国际实体传播知识与信息。

6）国际洪水倡议（IFI）

一个促进综合洪水管理方法的机构间倡议。旨在利用洪水和洪泛平原的优势，同时降低社会、环境和经济风险。合作伙伴：世界气象组织、联合国大学、国际水文科学协会和国际减灾战略。

7）国际水质倡议（IIWQ）

一个国际科学和政策合作倡议。旨在促进研究、生成和传播知识以及有效的和创新的政策，通过全面合作的方式迎接全球水质挑战，保障可持续发展的水安全。

8）国际共享含水层资源管理（ISARM）

一个专家网络倡议。旨在汇编世界跨界含水层清单，制定有关共享地下水资源管理的明智做法和指导工具。

9）国际泥沙倡议（ISI）

一个用于评估侵蚀程度和沉积物向海洋、湖泊或水库输移的倡议。旨在创建修复和保护地表水的整体方法，将科学与政策和管理需求紧密结合起来。

10）水资源综合管理（IWRM）

一种流域层级的水资源综合管理方式。作为更可持续管理水资源的一个基本要素，旨在产生长期社会、

ANNEX 4: IHP FLAGSHIP INITIATIVES

FRIEND-Water (Flow Regimes from International Experimental and Network Data)

An international research initiative that helps to set up regional networks for analyzing hydrological data through the exchange of data, knowledge and techniques at the regional level.

GRAPHIC (Groundwater Resources Assessment under the Pressures of Humanity and Climate Change)

A UNESCO-led project seeking to improve our understanding of how groundwater interacts within the global water cycle, how it supports human activity and ecosystems, and how it responds to the complex dual pressures of human activity and climate change.

G-WADI (Global Network on Water and Development Information in Arid Lands)

A global network on water resources management in arid and semi-arid zones whose primary aim is to build an effective global community to promote international and regional cooperation in the arid and semiarid areas.

HELP (Hydrology for the Environment, Life and Policy)

A new approach to integrated catchment management by building a framework for water law and policy experts, water resource managers and water scientists to work together on water-related problems.

IDI (International Drought Initiative)

The initiative aims at providing a platform for networking and dissemination of knowledge and information between international entities that are active working on droughts.

IFI (International Flood Initiative)

An interagency initiative promoting an integrated approach to flood management which takes advantage of the benefits of floods and the use of flood plains, while reducing social, environmental and economic risks. Partners: the World Meteorological Organization (WMO), the United Nations University (UNU), the International Association of Hydrological Sciences (IAHS) and the International Strategy for Disaster Reduction (ISDR).

IIWQ (International Initiative on Water Quality)

An initiative aimed at international scientific and policy cooperation to promote research, knowledge generation and dissemination, and effective and innovative policies to meet global water quality challenges in a holistic and collaborative manner towards ensuring water security for sustainable development.

ISARM (Internationally Shared Aquifer Resources Management)

An initiative to set up a network of specialists and experts to compile a world inventory of transboundary aquifers and to develop wise practices and guidance tools concerning shared groundwater resources management.

ISI (International Sediment Initiative)

An initiative to assess erosion and sediment transport to marine, lake or reservoir environments aimed at the creation of a holistic approach for the remediation and conservation of surface waters, closely linking science with policy and management needs.

IWRM (Integrated Water Resources Management)

Implementing IWRM at the river basin level is an essential element to managing water resources more sustainably, leading to long-term social, economic and environmental benefits.

JIIHP (Joint International Isotope Hydrology Programme)

A programme facilitating the integration of isotopes in hydrological practices through the development of tools, inclusion of isotope hydrology in university curricula and support to programmes in water resources using isotope techniques.

LaSII (Land Subsidence International Initiative)

The initiative enhances the scientific understanding and technical knowledge required to identify and characterize hazards related to natural and anthropogenic land-level lowering, especially due to groundwater resources overexploitation.

MAR (Managed Aquifer Recharge)

The initiative aims to expand water resources and improve water quality by promoting improved practices for

经济和环境效益。

11）国际同位素水文联合计划（JIIHP）

一个可以促进将同位素纳入水文实践的计划。旨在通过开发工具将同位素水文学纳入大学课程，支持在水资源项目中使用同位素技术。

12）地面沉降国际倡议（LaSII）

一个可以提高科学认知和技术知识的倡议。旨在识别和描述与自然和人为降低地面有关的危害，特别是由于地下水资源过度开采而造成的危害。

13）含水层人工补给（MAR）

一个可以旨在通过推广更好的含水层补给（存储和恢复）管理实践来增加水资源量、改善水质的倡议。

14）变潜在冲突为合作潜力（PCCP）

一个可以促进多层次和跨学科对话的项目。旨在促进有关共享水资源管理方面的和平、合作和发展。

15）城市水管理计划（UWMP）

一个产生方法、工具和指南的计划。旨在通过改善城市知识和分析城市水情，提出更有效的城市水管理战略。

16）全球水博物馆网络（WAMU-NET）

一个连接世界水博物馆、机构和站点的网络倡议。旨在展示人类与水及其自然的、文化的、有形的和无形的遗产之间多种形式的联系，通过分享典型经验和做法来支持新兴水文化的出现。

17）世界水文地质编图和评价计划（WHYMAP）

一个对全球水文地质信息进行收集、整理并可视化的倡议。旨在在全球水问题讨论中恰当表达地下水相关信息。

18）世界大河倡议（WLRI）

一个具有科学性质的倡议。旨在创建对世界大河现状和未来可能情形进行全面科学评估所需的知识库，基于良好实践制定创新战略，为同行学习和全球数据参考提供科学平台，实现大河可持续管理。

management of aquifer recharge (storage and recovery).

PCCP (From Potential Conflict to Cooperation Potential)

A project facilitating multi-level and interdisciplinary dialogues in order to foster peace, cooperation and development related to the management of shared water resources.

UWMP (Urban Water Management Programme)

A programme that generates approaches, tools and guidelines which will allow cities to improve their knowledge, as well as analysis of the urban water situation to draw up more effective urban water management strategies.

WAMU-NET (The Global Network of Water Museums)

The initiative connects water museums, institutions and sites worldwide that display diverse forms of humankind's connections with water and its natural, cultural, tangible and intangible heritage to support the emergence of a new water culture through sharing experiences and good practices.

WHYMAP (World Hydrogeological Map)

An initiative to collect, collate and visualize hydrogeological information at the global scale to convey groundwater-related information in a way appropriate for global discussion on water issues.

WLRI (World's Large Rivers Initiative)

This Initiative of scientific nature seeks to create the knowledge base required for a holistic scientific assessment of the status and possible future of the world's large rivers. For their sustainable management, it aims to develop innovative strategies based on good practices and serve as scientific platform for peer learning and global data reference.

5 变化理论图

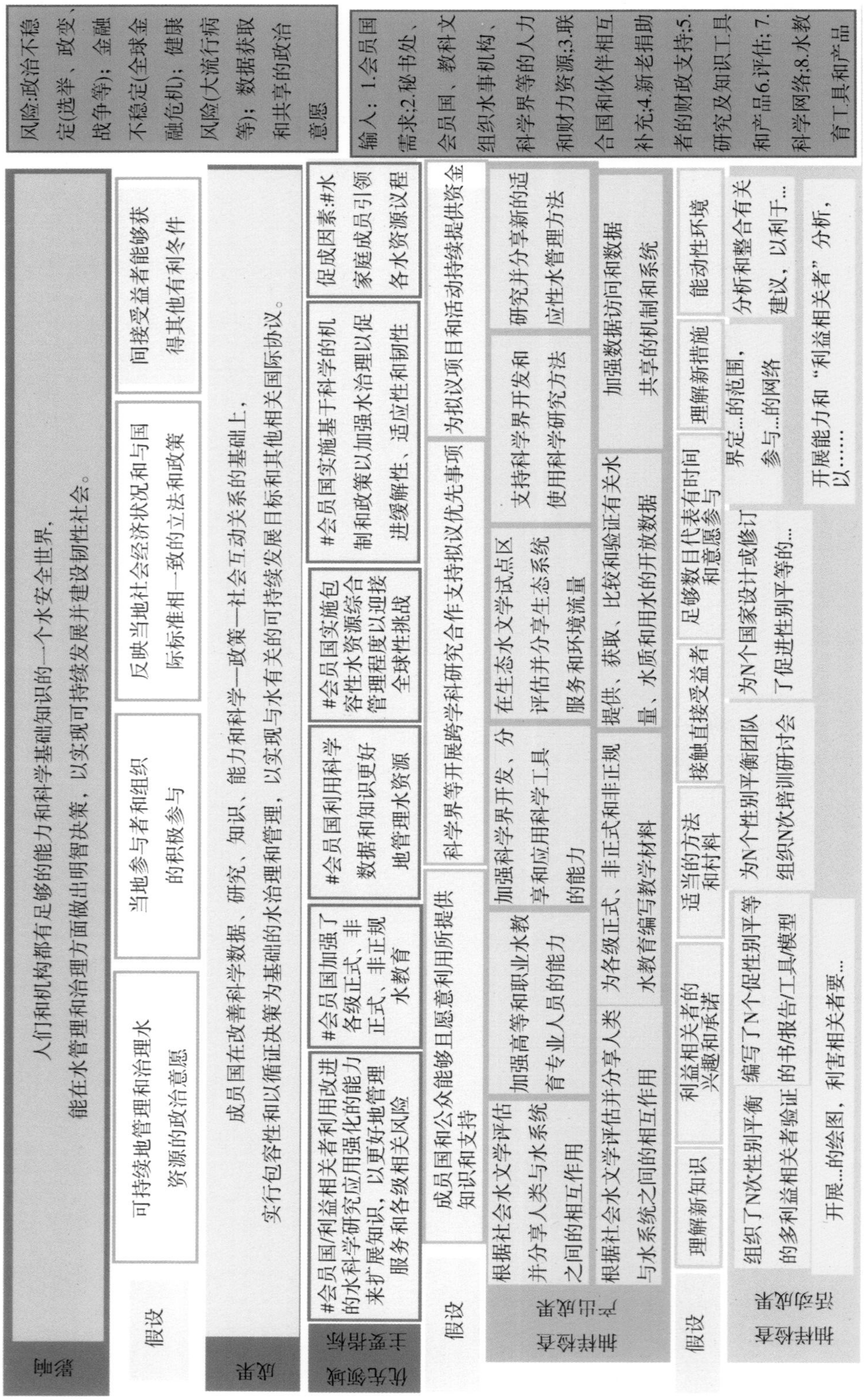

影响
人们和机构都有足够的能力和科学基础知识的一个水安全世界，
能在水管理和治理方面做出明智决策，以实现可持续发展并建设韧性社会。

假设
- 可持续地管理和治理水资源的政治意愿
- 当地参与者和组织的积极参与
- 反映当地社会经济状况和与国际标准相一致的立法和政策
- 间接受益者能够获得其他有利条件

成果
成员国在改善科学数据、研究、知识、能力和科学—政策—社会互动关系的基础上，
实行包容性和以循证决策为基础的水治理和管理，以实现与水有关的可持续发展目标和其他相关国际协议。

优先领域 主要指标
- #会员国/利益相关者利用改进的水科学研究应用强化的能力来扩展知识，以更好地管理服务和各级相关风险
- #会员国加强了各级正式、非正式、非正规水教育
- #会员国利用科学数据和知识更好地管理水资源
- #会员国实施包容性水资源综合管理程度以迎接全球性挑战
- #会员国实施基于科学的机制和政策以加强水治理以促进缓解性、适应性和韧性
- 促成因素:#水家庭成员引领各水资源议程

假设
- 成员国和公众能够且愿意利用所提供知识和支持
- 科学界等开展跨学科研究合作支持拟议优先事项
- 为拟议项目和活动持续提供资金

抽样检查 产出成果
- 根据社会水文学评估并分享人类与水系统之间的相互作用
- 加强高等和职业水教育专业人员的能力
- 加强科学界开发、分享和应用科学工具的能力
- 在生态水文学试点区评估并分享生态系统服务和环境流量
- 支持科学界开发和使用科学研究方法
- 研究并分享新的适应性水管理方法
- 根据社会水文学评估并分享人类与水系统之间的相互作用
- 为各级正式、非正式和非正规水教育编写教学材料
- 提供、获取、比较和验证有关水量、水质和用水的开放数据
- 加强数据访问和数据共享的机制和系统

假设
- 理解新知识
- 利益相关者的兴趣和承诺
- 适当的方法和材料
- 接触直接受益者
- 足够数目代表有时间和意愿参与
- 理解新措施
- 能动性环境

抽样检查 活动成果
- 组织了N次性别平衡的多利益相关者验证
- 编写了N个促性别平等的书/报告/工具/模型
- 为N个性别平衡团队组织N次培训研讨会
- 为N个国家设计或修订了促进性别平等的...
- 界定...的范围，参与...的网络
- 分析和整合有关建议，以利于...
- 开展...的绘图，利害相关者要...
- 开展能力和“利益相关者”分析，以……

风险:政治不稳定(选举、政变、战争等)；金融不稳定(全球金融危机)；健康风险(大流行病等)；数据获取和共享的政治意愿

输入：1.会员国需求;2.秘书处、会员国、教科文组织水事机构、科学界等的人力和财力资源;3.联合国和伙伴相互补充;4.新老捐助者的财政支持;5.研究及知识工具和产品6.评估; 7.科学网络;8.水教育工具和产品